THE
STEM CELL
HOPE

THE
STEM CELL
HOPE

How Stem Cell Medicine
Can Change Our Lives

ALICE PARK

HUDSON
STREET
PRESS

HUDSON STREET PRESS
Published by Penguin Group
Penguin Group (USA) Inc., 375 Hudson Street, New York, New York 10014, U.S.A. • Penguin
Group (Canada), 90 Eglinton Avenue East, Suite 700, Toronto, Ontario, Canada M4P 2Y3 (a
division of Pearson Penguin Canada Inc.) • Penguin Books Ltd., 80 Strand, London WC2R 0RL,
England • Penguin Ireland, 25 St. Stephen's Green, Dublin 2, Ireland (a division of Penguin Books
Ltd.) • Penguin Group (Australia), 250 Camberwell Road, Camberwell, Victoria 3124, Australia
(a division of Pearson Australia Group Pty. Ltd.) • Penguin Books India Pvt. Ltd., 11 Community
Centre, Panchsheel Park, New Delhi – 110 017, India • Penguin Books (NZ), 67 Apollo Drive,
Rosedale, Auckland 0632, New Zealand (a division of Pearson New Zealand Ltd.) • Penguin
Books (South Africa) (Pty.) Ltd., 24 Sturdee Avenue, Rosebank, Johannesburg 2196, South Africa

Penguin Books Ltd., Registered Offices: 80 Strand, London WC2R 0RL, England

First published by Hudson Street Press, a member of Penguin Group (USA) Inc.

First Printing, June 2011
10 9 8 7 6 5 4 3 2 1

REGISTERED TRADEMARK—MARCA REGISTRADA
HUDSON
STREET
PRESS
LIBRARY OF CONGRESS CATALOGING-IN-PUBLICATION DATA

Park, Alice, 1967-
    The stem cell hope : how stem cell medicine can change our lives/Alice Park.
        p. ; cm.
    Includes bibliographical references and index.
    ISBN 978-1-59463-078-1 (hardback)
    1. Stem cells—Research—Popular works.   2. Stem cells—Transplantation—Popular works.
I. Title
    [DNLM: 1. Stem Cell Research—Popular Works.   2. Stem Cell—Transplantation—Popular
Works. QU 325]
    QH588.S83P367 2011
    616'.02774—dc22

                                                                                    2011005022

Printed in the United States of America
Set in Bembo Std. Regular • Designed by Eve L. Kirch

PUBLISHER'S NOTE
Every effort has been made to ensure that the information contained in this book is complete and
accurate. However, neither the publisher nor the author is engaged in rendering professional advice
or services to the individual reader. The ideas, procedures, and suggestions contained in this book
are not intended as a substitute for consulting with your physician. All matters regarding your health
require medical supervision. Neither the author nor the publisher shall be liable or responsible for
any loss or damage allegedly arising from any information or suggestion in this book.

While the author has made every effort to provide accurate telephone numbers and Internet addresses
at the time of publication, neither the publisher nor the author assumes any responsibility for errors,
or for changes that occur after publication. Further, the publisher does not have any control over and
does not assume any responsibility for author or third-party Web sites or their content.

The art of progress is to preserve order amid change and to preserve change amid order.

—Alfred North Whitehead

CONTENTS

PREFACE

To the untrained eye the cells hugging the floor of the petri dish don't look any different from millions of their kind nurtured on thousands of culture plates in labs around the country. They clump together, smudging the clear plastic in random patterns, feeding off the liquid that serves as both home and nutrient for their laboratory existence. They spend most of their days in either incubated warmth or frozen animation, housed on ordinary shelves in ordinary labs. Even under the microscope, experts would be hard-pressed to tease out anything unusual or extraordinary about these cells. For newcomers like me, merely trying to detect their cloudy existence on the bottom of the dish is a chore.

After all that I have read and heard about embryonic stem cells, I have never seen them—images of them, yes, but not up close, in living form, swirling around in their culture medium. But I'm about to. In a cramped tissue culture room on the Harvard University campus, music from a radio beats in the background from a local pop station, and Sinisa Hrvatin, a third-year graduate student, can't help but sing along occasionally. It's the only room where the lab's principal investigator, leading stem cell scientist Douglas Melton, will allow a radio, because the work is so mundane. Hrvatin has logged so many hours in this walk-in

closet that he has memorized the lyrics of most of the songs. He is growing human embryonic stem cells—a labor-intensive, time-sapping,
365-day-a-year endeavor. The cells require daily care and maintenance,
and for several hours a day, he feeds them or picks out dead and aberrant colonies or breaks up the populations when they start to crowd the
plate. The constant attention means he doesn't go on vacation unless
he can find someone trustworthy to babysit his cellular charges, but he
doesn't mind—or says he doesn't. His goal is to find ways to morph these
stem cells into a very specific target—a pancreatic beta cell that makes
insulin.

Hrvatin turns to a small incubator on the floor that's no bigger than
a dorm refrigerator, pulls open the door, and picks out a stack of covered
plastic dishes, each filled with a thin layer of yellowish liquid. He holds one
plate up against a dark lab bench and points out the cloudy splotches smear-
ing the bottom of the dish. They look like water spots on an unpolished
glass but those, he says, are what I'm looking for: embryonic stem cells.

Like any other cells, these grow—dividing, splitting, duplicating
themselves over and over again into a living layer of copies. They represent a state in which everything is biologically possible, an expectant
phase before the march of development begins, before they embark on
the journey to turn themselves into the more than two hundred different types of cells in the body. They are the mother cells of the human
species, the ancestral core from which the cellular pedigree that is the
human body emerges.

And as such, unlike any culture of cells that has come before them,
these cells have the ability to sustain themselves in this pregnant state
of possibility indefinitely. As long as it has room to expand and food to
survive, a stem cell will self-renew, dutifully making two copies of itself.
One daughter cell will gradually age as it divides and develops, eventually expiring when it reaches the limit of its replicative abilities. The
other progeny makes a different choice, stubbornly refusing to mature,
instead retaining the remarkable ability to arrest itself in a state of perpetual embryonic youth.

This is what sets stem cells apart. These instructions for self-renewal make them the equivalent of a cellular fountain of youth. The molecular riches represented by this forever-young state are luring people to dream of hitting the stem cell jackpot. For patients, the cells represent the key not just to treatments, but even to cures for diseases that currently plague and perplex us—conditions such as diabetes, Parkinson's disease, spinal cord injury, and cancer, for which there are now only minimally effective treatments. For scientists, the cells are a living vein of unmined medical ore, clumps of precious information and windows into the intricately complex and mysterious choreography that is human development. How is the human body made? How do diseases occur? What switches lead normally developing cells to go astray, misdirected into disease? For ethicists and theologians, stem cells represent the ultimate temptation—the power to play God and redirect human development, and ultimately, human nature.

In science, all too rarely but every so often the convergence of creativity, innovation, technology, and technique yields an evolutionary step forward in our understanding of the natural world, our place in the universe, and even our own human and very mortal bodies. Vaccines, genetic engineering, gene therapy, and antibiotics all represent such platforms on which new insights in biomedicine were built. Immunizations and antibiotics forever tipped the odds in the age-old battle between man and infectious disease, finally affording the human species leverage against invisible but insidious pathogens from bacteria to viruses—but not before billions succumbed to the black plague, smallpox, and influenza, not to mention AIDS. The genetic era, which made possible the sequencing, identification, and manipulation of genes, bestowed upon us another gift—the opportunity to understand and even alter predetermined genetic fates, ushering in a new mastery of disease.

Stem cells are poised to do the same. With their ability to provide a limitless source of new cells to replace damaged and dysfunctional ones, they promise a still deeper command of the human condition. Already, embryonic stem cells gleaned from the very earliest days of development hold a tantalizing promise for addressing a myriad of different diseases.

So too do the recently isolated induced pluripotent stem (iPS) cells, a mouthful to pronounce but an intensely promising population of stem cells that appear to have all the properties of their embryonic cousins yet are generated without an embryo at all. Instead, iPS cells emerge from marinating a skin cell (or presumably any other adult cell) with a special genetic brew of just four early development factors. The process erases the skin cell's previous history and returns it to an embryonic state, from which it can become not just a skin cell again, but *any* cell in the body.

Scientists are engrossed in comparing the two populations of stem cells to figure out how similar they really are, and they are certain of one thing: Both groups represent a far more flexible and potent biological resource than the adult stem cells, which are found in the bone marrow, among other tissues, and are more limited in their transformative abilities—bone marrow stem cells, for example, are especially adept at creating more blood and immune cells, but less effective at making other tissues.

It's been only a decade since the first human embryonic stem cell line was generated in a small lab in Wisconsin, but stem cells are already transforming the way researchers think about disease, how clinicians think about treating their patients, and even how drug companies think about developing the next therapeutic blockbuster. It's now possible for researchers to grow in a lab dish the very cells that trigger disease—the defective nerves that fail to stimulate muscles in patients with amyotrophic lateral sclerosis (ALS), for example, or the weak light-sensing cells in the retina that lead to blindness in macular degeneration. Such recapitulation of a disease in a dish will eventually expose new routes to treating these disorders and make it possible to screen new drugs more efficiently and effectively than ever before.

Taken to a reasonable extreme, the development of these stem cells can lay bare a molecular and cellular universe of birth, development, death, and renewal that occurs over and over again with remarkable fidelity in every one of our body's cells.

And it's not just a blue sky scenario, but one that is already being executed in laboratories and biotech companies across the country. Using iPS technology, another group at Harvard has managed to grow and observe those motor neurons from ALS patients, and has already determined that the cause of their demise may be a toxic compound released by the glial cell, a neighboring nerve cell that normally provides nutrition and molecular support for the motor neurons. But in ALS patients, a mutated form of these cells ends up doing the opposite, destroying the nerves that reach out to muscle instead. It didn't take the researchers long to start screening drug compounds to see if the glial cell's damaging effects can be counteracted with a drug therapy.

This is the promise that draws, like a siren song of hope, a growing community of scientists, politicians, legislators, patients, families, and even average citizens to stem cells. They can all imagine a future in which doctors no longer have to guess at the black box of disease, a future when the processes that turn a diabetic patient's beta cells against him or her are no longer a mystery, or when the journey that a Parkinson's patient's brain neurons take from making healthy connections to causing erratic tremors is laid bare. That's the future they see. That's the future they hope for.

And it is the sheer force of this promise and potential that continues to propel the science of stem cell research forward, even in the face of strong moral and ethical opposition. Embryonic stem (ES) cells are removed from days-old embryos that cannot survive without them, and because of this, the field's very existence has become embroiled in the conflict over abortion, emerging as a matter of debate in Congress, the White House, and even the courts. For many, the destruction of embryos is too high a moral price to justify whatever good may come of research on them. In 1996, Congress legislated a ban on any government research that harmed or destroyed embryos, stymieing progress in stem cell studies. President George W. Bush also acted, issuing an executive order in 2001 that allowed for some government support of the field but essentially hampered any forward momentum for eight years. Even after President Obama removed these restrictions in 2009, the moral opposition reemerged, this time in the courts, with a lawsuit against the government claiming that any federal funding of embryonic stem cell research violated the law.

To what extent these actions were driven by political rather than ethical and philosophical concerns continues to be a matter of intense public, political, and scientific debate. But the fact is that in the short history of human stem cell research, the science has managed, somehow, to triumph. New revelations about the secrets of early development are overturning long-held views about the way biology works. Plumbing the inner workings of what gives stem cells their "stemness" led to the discovery of iPS cells, whose existence led to a radical rewriting of the rules of biology. iPS cells prove that cells are not restricted to a one-way journey from birth to development and death. That venture can actually be turned into a round-trip voyage—in a way that is both scientifically and ethically more straightforward than generating ES cells.

It was the recognition of this inexorable march of science that compelled President Obama, in March 2009, to lift the restrictions on federal funding of embryonic stem cell research in the United States. In expanding the number of human embryonic stem cell lines that the federal government can now fund, from a couple dozen to potentially hundreds, Obama noted that "the potential it offers is great," and "there is no finish line in science. The race is always with us—the urgent work of giving substance to hope and answering those many bedside prayers, of seeking a day when words like 'terminal' and 'incurable' are finally retired from our vocabulary."

How did we get to this point in scientific history? How did such a fundamentally profound shift in the very tenets of biology emerge, even in the face of public and political resistance? And where do we go from here? How do we take this newfound knowledge and transform it, safely and effectively, into the treatments and cures that researchers feel are imminent and that so many patients are desperately awaiting?

That tale is, at its heart, one of personal passion and commitment.

The series of extraordinary breakthroughs and equally spectacular break-downs behind the birth of a new discipline that many are referring to as "regenerative medicine" is not simply a chronicle of technical achievements, profound philosophical debates, and political rhetoric. It is the narrative of people, the story of those scientists, politicians, patients, and average citizens, and the passions and beliefs that inspired and compelled them. In this book, I highlight just a few of the dozens of individuals who have had a hand in defining the emerging field of stem cell science. The list is by no means exhaustive, but I hope, in the end, it is evocative of the zeal with which they have pursued this work, regardless of how they came to be connected with stem cells—from the scientist studying molecular mechanisms to the patient pushing for wider acceptance and funding of the field. This is their story.

ACKNOWLEDGMENTS

As I was reporting a cover story on stem cell research for *Time* in 2009, I was humbled, as I often am, by the creativity and force of will I encountered among the scientists I was interviewing. Not only were these researchers pursuing their work with the usual professional dedication to science and the pursuit of knowledge, they were doing so under an undeniable social and political cloud. In some cases, they kept their labs private, only disclosing their locations on a need-to-know basis in order to protect their students, their equipment, and their findings. This book would not have been possible without the cooperation of these pioneering individuals, who eagerly and without hesitation shared their time and insights with me over the course of numerous discussions. Dozens of researchers from the United States, the United Kingdom, Japan, and Korea did not hesitate to talk about their work and experiences for this book, and while not all of them are mentioned by name, without them, there would be no stem cells, and there would be no story.

In particular, I would like to recognize Douglas Melton, who opened his lab to me week after week, and carved out precious hours of his time to share not just his passion for stem cells, but the very personal and private reasons for why he became so enthralled with the field and all of its possibilities.

During one of our many conversations, Melton shared with me his belief in the strength of patient advocacy groups, and how these organizations have taken a leading role in defending and funding stem cell research. Valerie Estess, Robert Klein, Susan Solomon, and Jerry Zucker were all extremely generous with their experiences, and it was their dedication and that of others like them in the advocacy community that kept embryonic stem cell research alive when the federal government restricted its support of the field for nearly a decade. Without the financial support of these groups, many of the breakthroughs that are bringing stem cell techniques closer to treating patients would have taken years, if not decades, more to occur.

B. D. Colen, Paul Costello, Don Gibbons, and Jennifer O'Brien opened countless doors for me with introductions to the leading stem cell experts at their respective institutions. Their grasp of both the history and the latest developments in the field is inspiriting.

I am grateful to my editors at *Time* for running that story and recognizing the importance of stem cell science and its potential to change how we think about disease. Managing editor Rick Stengel was not only supportive of the idea of a book on the subject, but also passed along good advice on managing the overwhelming scope of the project.

My editor at Hudson Street Press, Caroline Sutton, took a chance on both me and a controversial scientific issue and for that I am eternally grateful. I was fully aware of how painful a book project could be, but from our initial meeting she made the entire process painless and improved the text immensely.

I am indebted to my friend John Powers of the *Boston Globe* for supplying invaluable advice and encouragement throughout the writing process, and for serving as my litmus test for how much detailed science a reader could take. He bravely read through early drafts and streamlined the copy as only a truly skilled journalist could.

Finally, to my family, who knew this book was in me even before I did. You have always been my foundation and my inspiration; words can't express how much I treasure your love and support.

1

A Tale of Two Fathers

November 4, 1991

Sam Melton was not a happy baby. For several days, the six-month-old's tiny body had been trying to tell him—and his parents—that something wasn't quite right. A generally healthy and easygoing infant, he had taken to lying listless in his crib, barely acknowledging his mother, father, or older sister when they tried to play with him or pick him up. Even more disturbing, he had stopped making eye contact with them, and a worrying red rash had bloomed across his bottom. He wasn't eating as much, but it was winter, in Boston. It was probably just a cold, his parents thought.

On this morning, something was different. After waging a frustrating battle with his own body for days, Sam had finally had enough.

He started to vomit. Or, more accurately, projectile vomit, shooting what little remained in his stomach straight across the room. Alarmed, his mother, Gail O'Keefe, immediately called the pediatrician. Hearing the symptoms, he wanted to make sure Sam wasn't getting dehydrated, so he suggested Pedialyte. Still flustered, O'Keefe realized after hanging up the phone that she had forgotten to mention another of Sam's symptoms: His breathing was getting faster and faster, as if he were gasping for air. But she gave him a spoonful of the medicine and started rocking him in her arms.

By midday, though, the syrup hadn't had much effect. Sam's complexion, normally a glowing, rosy pink, had darkened to a sickly gray. Nestled in his mother's arms, he hadn't kicked or stretched or fidgeted as he normally would. And try as she might, she couldn't get him to nurse. "I got really nervous," she says, "so I called the doctor again, and said I forgot to tell him about his respiratory rate." Unwilling to wait any longer, she also phoned her husband, Douglas Melton, and told him she was taking Sam to the doctor. She asked a friend to watch over her daughter, Emma, and rushed to the pediatrician's office. When they arrived, the doctor took one look at the baby and instructed O'Keefe to take her son to the emergency room—immediately. His nurse, who had overseen Sam's pediatric visits since he was born, started to cry; she wasn't sure the baby would make it.

Melton met his wife and son at Children's Hospital Boston, one of the leading pediatric medical institutes in the country. Not knowing what was wrong with Sam, doctors launched a series of tests that would stretch the Meltons' growing anxiety well into the early evening. Blood tests, some poking and prodding, and more poking and prodding provided no answers. Was it a gut issue? Something Sam had eaten? A reaction to something he had touched? Complications from the flu? At one point, they gave the baby a barium enema to clear his digestive tract, suspecting that he had a blockage in his intestine. It was their best guess as to what was ailing the six-month-old, given that he had been vomiting a few hours before.

But the enema was clear, and the emergency team moved on to ever scarier possibilities. The situation was getting more and more desperate— Sam remained lethargic, and O'Keefe and Melton were frantic that their baby son might not survive the day. "He was about to expire," says Melton, who two decades later still finds the events of that day difficult to discuss. The very life seemed to be seeping out of Sam. There was talk about doing a spinal tap to rule out meningitis, but like all the other tests they had been performing over the course of eight hours, it was only a guess. Time was running out.

And then it happened. Sam's ailing body mustered one final cry for help. He was losing the fight, and there was only one way that his cells could get the help they needed.

He peed.

Lying on the examination table, Sam relieved himself in a puddle that proved far more informative than any of the high-tech tests he had endured that day. As a liquid deposit for the by-products of the body's metabolism, urine generally carries waste without much diagnostic value. But because it concentrates the body's biochemical castaways, it can also serve as an indicator of what is off balance. And in Sam's case, it was his pancreas. Rather than simply wiping the puddle away, a nurse noticed that the urine smelled sweet and decided to stick a test strip into the mess on the table—just in case. What she found changed the Meltons' lives—and quite possibly the course of stem cell science.

Sam's urine was full of ketones. For hours, and possibly days, his body had come dangerously close to shutting down in a desperate and ultimately futile search for the insulin that it needed to keep all of his metabolic functions going. When his cells couldn't find any, his body began burning up its own fat stores and releasing the acidic ketones as a by-product. It wasn't that he hadn't been fed—he had. But his body was starving. He was consuming glucose but his pancreas wasn't making enough insulin to break it down from the milk he was drinking.

Sam was diabetic.

In retrospect, it's such an obvious diagnosis, and the doctors at a leading children's hospital might have considered the possibility that Sam was suffering from the buildup of ketones in his blood much sooner. But diabetes doesn't normally show up in infants so young. In fact, at the time, Sam was the youngest baby on record diagnosed with the disease in the hospital's 122-year history.

Once the diagnosis was confirmed, the medical team acted fast. Nurses inserted an IV with insulin into his tiny arm, and, says O'Keefe, "Within twenty-four hours, in intensive care, he was on his hands and knees, and he was nursing. He was a baby again. It was just the most amazing thing in the world—a little insulin. It was an incredibly dramatic presentation of this disease that we knew nothing about."

But she and her husband quickly learned. The diagnosis certainly gave Sam a very different life from the one his parents had envisioned. He would always be dependent on outside sources of insulin hormone to compensate for what his body could not make. He would have to keep careful track of what he ate and how much energy he expended, to make sure his body was getting the right amount of glucose at the right time.

The diagnosis changed Melton's and O'Keefe's lives as well. Melton had been trained as a biologist, and while he was a scientist, not a physician, he knew what lay ahead for his baby son: years of constant finger pricks to check the sugar in his blood, and repeated insulin shots to mop up the excess that his body could no longer process.

O'Keefe became, in essence, her son's pancreas, developing almost a sixth sense for when he needed more insulin—in the middle of the night, or after an exhausting afternoon of soccer practice—or for when his insulin levels had climbed too high and he needed a candy bar to soak up the overflow.

The week that Sam was diagnosed, Melton put his research duties in the lab on hold and stayed home with his wife, daughter, and son. Those days following Sam's diagnosis, he says, were a "horrific time, not one I like to remember." A professor in developmental biology at Harvard University, Melton was already well known and respected for his work on early amphibian development. Frog embryology, not human development, was his specialty. But after Sam returned home from the hospital, Melton was haunted by the constant blood checks and insulin injections in Sam's future. In his worst moments, he confronted the possibility that even with the most vigilant care, Sam could have severe health problems as he reached adolescence and adulthood. He knew that a bottomed-out blood sugar could lead to blackouts, organ failure, even blindness or amputations. He considered all of this and came to a realization. "I was not just going to sit around," he tells me as he relives that painful day as we talk in his office on the Harvard campus. "I was going to do something."

If a cure didn't exist for type 1 diabetics like Sam, Melton and O'Keefe decided, then Melton would find one. Or at least try.

As radically different as amphibians are from humans, Melton's shift wasn't as preposterous as it initially seemed. Both involved a similar appreciation for the intricate choreography of development, an understanding of what drives a cell to turn on certain genes and turn off others as it maps out its fate. Compared to the changes his son would face, Melton determined, the career turnaround was not such a sacrifice.

In type 1 diabetics, the beta cells, the specialized factories in the pancreas designed to dole out insulin in response to glucose levels in the blood, no longer function. It's still not clear why, but the leading hypothesis is that the body's own immune forces somehow see these cells as foreign, and launch an all-out attack on the unsuspecting tissue. Without a built-in way of distributing insulin when it's needed, type 1 diabetics need to provide the hormone from an outside source and, for all intents and purposes, find a substitute for their pancreas.

Sam would spend the rest of his life looking at food in a special way. Every bite would be a calculation, a mental tally of calories and glucose content that would need to be counteracted with a shot of insulin. An insulin pump—a device inserted just under the skin in the abdomen, hips, or thigh that delivers insulin continuously—can make the delivery easier, but as smart as the machine is, it still cannot calculate, as a normal pancreas does, when and how much insulin a diabetic needs at any given time. So it's still up to the wearer to decide when, and what dose, to dispense.

It wasn't a future that Melton could simply accept for his son. "I had the facility to do something about it. I was trained so that I could change my career path." So, he says, he gave up working on something he was familiar with and decided instead to take on something he knew nothing about. He then quickly counters any suggestion that his decision was a heroic or even exceptional act. "Remember the larger context. I was a tenured professor at one of the nation's best universities. I was a Howard Hughes Medical Institute investigator," he says. "If I couldn't take the risk, who could? And it's not like I switched from being a reporter to be a nuclear physicist. I was working on how animals develop, using frogs, and I switched to work on how mammals make a pancreas."

While there is ample literature on the anatomy of the pancreas—O'Keefe recalls her husband lugging home probably every textbook on the organ from the medical school library—there is precious little on its development, much less specifics on generating the different types of pancreatic cells. Melton, it appeared, had quite a mountain to climb.

But returning home every day, he was reminded of why he was making the journey. As a frog biologist, Melton was a scientist who happened to be a father. When he returned to the lab after his week's hiatus, he was a father who happened to be a scientist. And as a father, he explained in an emotional meeting with his lab members, he was redirecting his research focus away from amphibians and toward the human pancreas. His lab now had a mission, and that mission was to find a cure for diabetes. To find a cure for Sam.

September 2001

It was the week after the most stunning and tragic terrorist attack on the United States, and like most American families, the Kleins, though still stunned, were attempting to reestablish the rhythm of their lives. Robert Klein, a successful real estate developer in Northern California, had watched the horrifying events of 9/11 unfold on television from Europe, where he was traveling with his daughter on a long-promised trip to celebrate her high school graduation.

Back home in Portola Valley, his youngest son, Jordan, was starting seventh grade at a new middle school. Anxious, excited, apprehensive, and expectant, the eleven-year-old hadn't been feeling quite right. He was losing weight, and inexplicably experienced intense stomach pains that would suck the breath right out of him. His mother noticed that he was also drinking copious amounts of water and going to the bathroom constantly. She recognized the symptoms, which she'd seen in one of Jordan's friends who had been recently diagnosed with diabetes, and she suspected that her son, too, might have the metabolic disorder. But it was too difficult a reality to face quite yet. Maybe Jordan was just nervous about the new school, and the stress was triggering his headaches and stomachaches and fueling his frequent trips to the bathroom. Maybe.

It didn't take long for her suspicions to be confirmed. Sitting in an English class and trying hard to pay attention, Jordan was seized by a breath-robbing punch of pain that knocked him off his chair. With his entire body on fire, feeling as if all of his blood had turned instantaneously into caustic battery acid, Jordan saw nothing but black.

His mother rushed to the school and took him to the hospital. After a series of tests, the doctors told her what had been ailing Jordan.

Klein had been calling in every day from overseas to check in, and was completely unprepared for what his wife told him after the harrowing trip to the hospital. "We know what's wrong with Jordan," she said to her husband over the phone. "He has juvenile diabetes."

The next thing Klein heard over the receiver was a chilling scream— a raw, wrenching outburst that left him feeling as if he'd been physically struck in the gut.

It was Jordan.

His mother had planned on discussing the diagnosis first with his father before breaking the news to him, so had spared her son the news initially. She had not realized that he was listening.

"He screamed one of those uncontrolled kind of primitive screams that . . . represents total fear, as if something was going to destroy you in the middle of the night," Klein tells me, taking long pauses as he struggles to contain the painful memories our conversation brings up. "If you can imagine the worst fear of your life . . . and you hear . . . this was a primitive scream of terror. He was terrified that his life was being taken away from him. Because he knew what the disease could do."

Unlike Sam, who was only six months old at his diagnosis, Jordan was fully aware of what juvenile diabetes, or type 1 diabetes, could do. For years, he had watched a Cub Scout friend deal with the disease. He knew all about the insulin injections that kept his friend alive, as well as the myriad potential dangers looming in his buddy's future—the threat that the ever-undulating levels of glucose posed to his kidneys, his heart, his eyesight, even his arms, legs, fingers, and toes as they began to compromise his body's ability to circulate blood and nutrients to his limbs. Jordan was well aware of all of this because he and his father had talked about it.

But now it was happening to him. That future of potential kidney damage, blindness, even amputation was no longer someone else's to worry about. It was his. And suddenly everything was different.

Five thousand miles away, Klein could not hug his son, could not provide a shoulder to lean on or the physically soothing presence that only a parent can give a child. He had only his voice. "I think I said to him . . . I'll be there for you," Klein says, tears beginning to well in his eyes. "We'll get through this together."

Getting to his son, however, would prove to be a problem. For days, air traffic into the United States had been completely shut down, and still, a week later, flights were booked to capacity as airlines struggled to accommodate the pent-up demand and scrambled to implement new security measures. For two days, Klein worked the phones from his hotel, desperate to find a way to reach his son before the following week, when he and his daughter had originally been scheduled to leave. When that failed, he turned to research. If he couldn't physically be there for Jordan, he would become his son's ultimate source for all things diabetes. Over the next week, he nearly wore out the fax machine at the hotel, tallying up about a thousand dollars in fees when his office beamed over as many pages from books on the disease as they could. Hungry for any information that might help Jordan, Klein read them all. At the time, those textbook sources still did not contain any mention of stem cells and how they might be used to treat or even cure diabetes; the science was still too new.

Klein had yet to fully appreciate stem cells, either for their scientific potential in alleviating diseases such as diabetes or for the formidable political challenges they raised. But it wouldn't take long. Joining the Juvenile Diabetes Research Foundation (JDRF) as a board member, he could hardly miss the growing excitement over the mother of all cells. These cells, many scientists were telling the foundation's members, could finally break the back of diabetes, and crack open the door to a cure. He couldn't help but get caught up in the hope the cells represented. If embryonic stem cells could be grown up and manipulated to form insulin-producing cells, then these cells might be transplanted into patients like Jordan to save them from the tyranny of needles and glucose monitors and the always looming threat of complications.

The scientific challenges to making that happen were certainly daunting, and the person tempering these researchers' enthusiasm more than anyone or anything else was the president of the United States. Just a month before Jordan's diagnosis, George W. Bush had taken the unprecedented step of restricting federal funding of the very embryonic stem cell studies that many scientists were beginning to believe would be critical to finding new therapies for diabetics. Citing concerns over the only available source of the cells at the time—human embryos—Bush limited taxpayer support to a small number of stem cell lines already in existence, and prohibited government dollars from funding any new lines. If these cells were to hold a cure for Jordan, Klein knew, there would have to be more lines.

As a successful real estate developer in California's booming housing market, Klein did not possess the scientific skills to push the biological limits of what stem cells could do, as Melton had begun to do nearly a decade before. But he did have the ability to address the challenges facing the field outside of the lab—in Congress, on the streets, and in the minds of the American people. With the debate over the embryonic origins of the stem cells threatening to overwhelm and suffocate their scientific potential, Klein took on the task of changing the public image of embryonic stem cells. Without government funding of the field, he knew, progress in finding cures for patients like Jordan would slow to a crawl. But convincing legislators to change the restrictive policies would take too long. Researchers needed more lines to fully explore the potential of these cells, and if the federal government wasn't willing to step in and support the work, Klein thought, then perhaps a state legislature could. And what better state than California, with its embodiment of the pioneer spirit and as home to some of the field's most talented stem cell scientists, to forge the way? No state, however, had ever taken on the role of single-handedly financing a biomedical science. Public works projects, yes. Schools and highways and utilities, yes. But backing stem cell research would be an unprecedented challenge, as Klein would find out.

2

It Began with Dolly

Shepherds have a saying about sheep: They spend most of their days dreaming up new ways to die.

So it must have come as somewhat of a surprise to the long-suffering shepherds in the lowland Scottish village of Roslin that a starry-eyed scientist had actually chosen their finicky animals as a model for learning about early development and reproduction.

Imagine their skepticism, then, when they heard that this misguided soul was actually attempting to turn their woolly charges into a living pharmaceutical factory to churn out potentially lifesaving and disease-curing compounds in their milk. Daft, they thought. Completely and utterly daft.

Ian Wilmut now smiles when he recalls those days B.D.—Before Dolly. Bespectacled, with soft tufts of reddish hair that float around his balding pate and soften his chin in a fuzzy beard, the embryologist hardly seems the type to shake up decades of entrenched biological wisdom. Or, for that matter, to incite the passionate protests—and wrath—of scores of religious and conservative thinkers who were alarmed at the experiments he had conducted in a small barn on the Roslin Institute campus's farm near Edinburgh. Introspective, Wilmut rarely raises his voice above a murmur, and in public, particularly during discussions on the endless parade of panels to which he is invited to speak about Dolly, he tends to sink his chin into his neck and look downward, as if in a perpetual state of abstraction. Even now, a dozen years after Dolly's birth, Wilmut seems less comfortable in the spotlight than he would be in the Scottish countryside that came to define him.

Like it or not, Dolly, his crowning professional achievement, would close the door on the privacy of life as a basic scientist that Wilmut had come to cherish. Because he ended up doing far more than finding a way to supercharge sheep's milk. He cloned the animal itself, making a genetic copy of an adult ewe from another ewe's mammary cells. The result was not only the first sheep ever to be cloned, but the first mammal to be cloned.

He named her Dolly, after Dolly Parton—for obvious reasons, he says, given the lamb's unique provenance—and she became perhaps the most photographed and coddled sheep in history.

At the time of Dolly's birth, it wasn't obvious that she, or cloning, would have any role to play in stem cell science at all, but with every year that passes, her legacy in the field becomes clearer. For in creating Dolly, Wilmut had, in essence, bestowed upon us two very significant gifts—the discovery that adult cells could be reprogrammed, and the key to unlocking the potential of human embryonic stem cells.

★ ★ ★

"When I think about the significance of the Dolly experiment, it's the stem cells, which in turn influenced people's thinking, in making people think differently and much more ambitiously about what can be done with cells, that is by far the most important outcome," Wilmut reflects. Still in Scotland, he has moved from bucolic Roslin to the capital, where he now directs the Centre for Regenerative Medicine at the University of Edinburgh—a facility and program born, in large part, out of his work in cloning.

It took time, however, for this broader impact of Dolly's birth to be felt among both scientists and the public at large. When Wilmut first announced Dolly's existence in a paper published in the journal *Nature* in 1997, it was understandably the cloning part of her provenance that everyone noticed. What Dolly proved was that a fully developed cell such as one from the mammary could not only live a second life as something else entirely—as a nerve cell or a muscle cell, for example—but could serve as the seed for an entirely new animal.

Until Dolly entered the world in July 1996, scientists were convinced that mammals, occupying the highest rungs of the evolutionary ladder, developed along a one-way trajectory that was likely both permanent and irreversible. As with a rolling odometer, once the countdown of development began, there was no turning back the biological clock, no rewriting the genetic legacy of a cell in a biological do-over.

And there was ample reason to believe this. It wouldn't do for development to be so unstable as to flip back and forth between different stages of maturity, for cells to be toggling between varying identities. You can't have a brain cell deciding to act like a heart cell, for instance. The road to adulthood had to be a one-way street in order to keep the task of growing from embryo to adult under control. Along the way, it appeared that certain checks to this directional progression were written into the DNA of cells, presumably triggered as a cell read its genetic code and proceeded along its preordained fate from embryonic cell to muscle cell or a nerve nexus in the brain.

At least, that was the conventional wisdom in the 1970s, when Wilmut first arrived to set up his labs in the hills of Roslin. Born in Hampton Lucy, a small town near Warwick, England, to teachers, Wilmut was only an average student in school, with dreams of becoming a farmer. By the time he had graduated from the University of Nottingham, however, he had become enthralled with the earliest stages of life, the intricate choreography of molecular steps that turn one cell—the fertilized egg—into two cells, then four and then eight and eventually into an entire multicelled animal.

By then, the young embryologist had also fallen under Scotland's spell. In *After Dolly*, his own account of Dolly's birth, Wilmut acknowledges a fondness for the "traditional Scottish pursuits of drinking whiskey, walking, and staring into space." During Roslin's long, blustery winters—which also happen to be mating season for the sheep that became the object of his professional attentions—Wilmut also passed the time perfecting his skills in the Scottish sport of curling.

The Roslin Institute was, and still is, a leading animal science research center, and its adjacent Dryden farm is home to a menagerie of farm animals, including chickens, pigs, cows, and, owing to its location, sheep. Soon after Wilmut began his career as an embryologist there, however, drastic cuts by the British government to science research budgets (Roslin faced a two-thirds shrinkage in its funds) threatened to shatter his Scottish reverie. Roslin's science-for-the-sake-of-science mission, the staff now learned, had to be laced with a more pecuniary purpose—preferably one with a commercially bright future.

For agricultural scientists like Wilmut, that meant that his basic pursuits, inspired by those bouts of staring off into space, needed to have a tangible biotechnology outcome—and in 1982, his director abruptly ended Wilmut's seemingly unprofitable work on embryos and presented him with a new project.

As a genetic engineer, Wilmut would have the task of tinkering with the sheep genome and endowing the animals with the ability to produce a pharmacopoeia of milk laden with antibiotics, clotting proteins useful for treating hemophilia, or even alpha-1-antitrypsin (AAT), a protein that keeps destruction of lung tissue in check. These would then be purified and turned into a pharmaceutical product to treat human patients with cystic fibrosis or other lung diseases.

It was tedious and painstaking work. Around the same time, a biochemist at Stanford, Paul Berg, was beginning the work on splicing and recombining DNA that would earn him a Nobel Prize and change the way scientists like Wilmut would conduct their genetic experiments. But recombinant DNA was not yet as ubiquitous as it is in today's labs, and his attempts at genetic engineering were, Wilmut says, slow going.

After several months of trying, however, he finally succeeded in generating a "pharmed" sheep, named Tracy, who dutifully churned out AAT, normally available only from human plasma. Because AAT was often in limited supply, Roslin and its financial partner, the biotech company PPL Therapeutics, were hoping sheep like Tracy would provide a far more reliable and scalable source of the protein. With sufficient amounts of AAT-laden plasma, more patients suffering from emphysema or cystic fibrosis–related respiratory disorders could protect their lung tissue from deteriorating, and the company was hoping to corner the market on the "pharming" method. And as an AAT factory, Tracy put her fellow genetically engineered barn mates to shame—with every liter of milk, she produced thirty-five grams of the precious protein, while four of the other genetically altered sheep together could manage to eke out only a gram.

Tracy, however, was one of a thousand embryos that Wilmut had painstakingly injected with the gene for AAT. Adding genes to an embryo's genome is less a feat of scientific precision than a game of genetic roulette, with equally unpredictable results. It didn't help matters that early-stage mammalian embryos actually have two nuclei packed with genetic material—one from the egg and one from the sperm—which fuse to become a coherent bundle of densely packed chromosomes. Gene splicers like Wilmut had no way of telling which nucleus was which, and ended up targeting either one, although the egg's pronucleus is often preferred, since it is physically larger and possibly more stable. Of course, not all of the embryos welcomed the genetic intrusion, and at the time, only about 1 percent of them saddled with such genetic guests actually yielded a live animal with a functional version of that gene. And even then, it was a roll of the genetic dice as to how well that particular hybrid animal could access the gene and produce the protein in question. In Wilmut's case, it was a process made more frustrating, he readily admits, by a natural tremor in his hands that made it nearly impossible for him to conduct the precise manipulations needed to first position and then inject the embryos under a microscope using a micropipette.

After bumbling his way through injecting thousands of embryos, Wilmut figured there had to be a better way.

★ ★ ★

Indeed, there was. If the problem with gene targeting was its unpredictability (Wilmut could never tell which embryos had successfully integrated the gene so the resulting animal could produce the encoded gene's protein in generous amounts), then the solution was to find a way of generating embryos that would readily and reliably welcome such genetic additions.

The most stable way to generate such embryos is to clone them. Once an embryo contains the right gene in the right place in its genome, then making copies of that embryo to generate similarly altered animals is the most straightforward route to such a consistent population of animals. And that wasn't actually such a far-fetched scheme. Wilmut, like any well-trained developmental biologist at the time, was familiar with the seminal experiments that his countryman, John Gurdon, had conducted on cloning in the 1960s.

If you scan the acres of articles written in the press about stem cells, from the breathless reports of the latest breakthroughs to the politically charged invectives denigrating the temerity of scientists who dare to fool with Mother Nature, you might find only passing mention of John Gurdon, if at all. Search the scientific literature, however, and a 1958 paper by Sir John Gurdon (he was knighted by the Queen in 1995 for his scientific accomplishments) is one of the most frequently cited references in the stem cell field—because it was Gurdon who actually cloned the first animal, a frog, from a fully developed cell. And in doing so, he set the stage for Wilmut to think about using a far older cell to create Dolly. Gurdon had answered a critical question in developmental biology: whether cells retain their embryonic ability to become any type of cell in the body, or whether, in the course of their development, they permanently lose or silence all developmental journeys except for the one they choose. (Several decades later, Gurdon would accept a young American graduate student who was similarly intrigued by the morphing abilities of cells and inspire him to think about the power of regenerative biology—Douglas Melton.)

A "man of independent means," as Wilmut and his fellow Brits describe him, Gurdon took an atypical journey to the lab bench that did not follow the usual route from early curiosity and precociousness to a more mature inquisitiveness. Instead, as a fifteen-year-old high school student in a privileged school, Gurdon was such a poor science student that his teacher at the time felt compelled to warn him—and his parents—away from the field. "Gurdon has ideas of becoming a scientist," the instructor wrote in one report. "This would be quite ridiculous. If he can't learn simple biological facts, he has no chance at all of going on with science. It would be a sheer waste of time both on his part and those who have to teach him." Amused by the teacher's obvious frustration, Gurdon keeps the withering assessment on his desk to this day, and happily reads it to anyone who asks about his scientific start. It wasn't that science didn't interest him, he explains, but that he learned without textbooks. Classes were all about memorizing facts instead of thinking about the principles behind them. "That terminated my science career immediately at age fifteen," he says with a dry laugh.

But undeterred by his schoolmaster's warning, the young Gurdon continued to immerse himself in weighty tomes on entomology, buying them himself and devouring every word. When he applied to Oxford, however, it was Michael Fischberg, a professor in developmental biology, not entomology, who took an interest in Gurdon and offered him a place in the lab.

Fischberg was studying frog embryos.

A few years before Gurdon began his graduate studies, a pair of American researchers in Pennsylvania had begun stretching conventional ideas of early development with innovative experiments in which they switched out the genetic material of reproductive cells. Robert

Briggs and Thomas King, working at a cancer institute in Philadelphia, took the nucleus of one frog cell and transferred it to a frog egg that had been emptied of its own nucleus. Miraculously, the pair reported that the resulting embryo actually began to divide—faithfully copying the DNA, presumably from the donor cell, since the egg no longer contained a nucleus, squeezing the new genetic material into first two, then ever more new cells. For all intents and purposes, the embryo with the swapped-out nucleus acted like a normally developing one.

It was a transplantation experiment like none that had been attempted before. The cell that Briggs and King transferred into the egg had come from a blastula, a hollow ball of several hundred cells that forms in frogs just days after fertilization. For any cell, the nucleus serves as its heart and soul. Packed with miles of genes twisted into tightly twined ropes of chromosomes, the nucleus houses the DNA that gives a cell its identity as well as instructions for how to survive—where to go to find the optimal nurturing conditions, and how to seek out proper nourishment to grow and flourish. Without its nucleus, a cell is an inert amalgam of cytoplasm and organelles without a purpose.

All cells begin as equals in the fertilized egg. Once sperm and egg fuse, those first cleavages create cells that retain a similar ability to contribute to all of the tissues that eventually emerge in a fully formed adult—a trait known as pluripotency. As cells mature, they gradually lose this ability and become increasingly multipotent or unipotent (specialized to become a certain type of cell able to perform a designated function, such as, say, conducting electrical impulses or producing insulin). But how does this gradual differentiation happen? Do cells embarking on a path to becoming a heart cell permanently lose their ability to become a skin cell? Is the differentiation process permanent and irreversible, meaning that a heart cell could never become a skin cell? And if so, at what point does this specialization occur?

Briggs and King thought they had the answer. While the blastula cells resulted in viable new frog embryos when transplanted into a nucleus-free egg, cells even one day older did not—no viable embryos, let alone tadpoles. They concluded, reasonably, that some change occurs just after the blastula is formed that wipes away the very early cells' unique pluripotency. Something, they decided, was permanently altering differentiated cells to prevent them from ever turning back and becoming pluripotent again.

Fischberg was interested enough by the theory to encourage his new charge to replicate the work, but with a different species of frog, *Xenopus laevis*, which he happened to have in good supply in his laboratory at the time. With the right hormonal encouragement, *Xenopus* had the added advantage of shedding eggs year-round, giving Gurdon the opportunity to run his experiments in generation after generation, whereas the leopard frogs ovulated only once a year.

Rather than focus on the features of the transplanted cell and its ability to revert back to an embryonic state, Gurdon, still young and inquisitive, and perhaps a bit naive, turned his attention to the egg. At its heart, he realized, nuclear transplantation was about the egg's ability to reprogram cells and establish a biological blank slate. After all, when sperm fertilizes an egg, the egg flips the ultimate identity switch by erasing the sperm's genetic profile and remaking it into one inextricably fused and integrated with its own. But how universal was this process? Briggs and King had proven that it was not unique to sperm, but Gurdon wasn't quite ready to accept that the process would not work for any cell, no matter how mature. Given the right conditions, he thought, the egg might be capable of exerting its genetic morphing powers on any specialized cell.

"Remember that the main point was to test the proposition that [differentiated] cells do or do not have the same genes," he says. "In the 1950s, one didn't know that." To find out, he methodically transferred increasingly differentiated cells, from late blastula and even further advanced gastrula stages, which represent the next phases of embryonic development, into recipient eggs. Much to his surprise, he says, "It started working in about a year. And within two years, it was already clear that it was working pretty well."

Just six years after Briggs and King had sounded the death knell for transplanting the nuclei of more mature donor cells, Gurdon published a one-column description of his results in the journal *Nature* in February 1958, contradicting their claim that only early embryonic cells could generate new frogs. He noted that transplanting even older cells resulted in viable embryos that "appear normal at various stages of development up to the feeding tadpole."

By 1966, Gurdon had extended the work to the most mature cells transplanted at the time—intestinal cells from tadpoles. The once inept science student achieved remarkable success in generating tubfuls of tadpoles. In Gurdon's hands, even fully differentiated cells already specialized to function in a frog's intestinal lining were fair game for the egg to reprogram.

"My results of course did not agree with Briggs and King, and as I was a graduate student, and they were well established scientists and much respected, it was obvious that the graduate student was likely the one who got it wrong," says Gurdon with his characteristic self-deprecating humor.

In the first hint of the careful and thorough scientist he was becoming, however, Gurdon had integrated proof of his work into the experiment. Exploiting newly developed ways of marking the nuclei of cells that Fischberg had just developed, he was able to prove that the nuclei of the donor intestinal cell had indeed given rise to new tadpoles.

The implications were enormous. It meant that rather than shutting off or losing genes as they matured, cells retained their entire complement of genes, selectively silencing or activating those that it needed to perform its specialized function. Every cell, then, could theoretically be made young, or in biological terms, embryonic, again. "The main conclusion from our studies was that every cell has a complete set of genes," he says. "And if that is true, then in principle, you can derive an embryo or stem cells from any cell of the body. You couldn't do that if genes are lost or permanently inactivated, because you could never take adult cells and make them then go backwards to being embryo cells—that could not work."

Making cells go backward—that was what Wilmut, pondering his dilemma of improving the efficiency of his genetic targeting nearly three decades later, decided might provide a solution to his intractable problem of inefficiency with gene targeting. Yet while it seemed possible that cells could be reprogrammed in amphibians, there was no evidence to suggest that the same would be true in mammals such as sheep.

★ ★ ★

Although the word never appears in Gurdon's papers describing his work, what he actually did was produce the world's first animal clones. When I ask him about why he never referred to his experiments as cloning, he is genuinely perplexed. "I didn't really think of it that way," he says. "I just happened to prefer *nuclear transfer* because that's what it was, so that's the term I used."

His scientific exactitude notwithstanding, the rest of the world had quickly caught on to the implications of Gurdon's experiments. Walter Cronkite made the trip across the Atlantic to interview the young graduate student, and asked the inevitable question, using the *C* word: When did he think the same cloning process would move from frogs to mammals and eventually to people? Gurdon's cheeky answer? "No idea, but somewhere between ten years and a hundred years."

As it turned out, it would take three decades for Wilmut to consider attempting the feat in his sheep. Others had tried, using mouse cells and embryos, and one of Gurdon's own students attempted to clone a rabbit, but their failures suggested the process would not work in mammals.

Even Wilmut had no interest in generating a herd of cloned sheep simply for the sake of cloning. What he was after was a rich population of cells that could grow well in a petri dish, proliferating at a reliable and steady pace once they had been tweaked to contain specific genes for therapeutically useful proteins. "The thing you need is a cell that can grow for a long time in the lab," he says. "There are lots of steps to genetically modifying a cell, so once you put in a gene and find the ones that have the right change, then you have to grow them, and they have to be working cells that can grow for a long time." Wilmut figured these cells could then be cloned, using Gurdon's technique, and presto! Problem solved. If, as Gurdon had proved, the developmental clock on cells could be turned back and they could be reprogrammed to become entirely new beings, then why not turn genetically altered sheep cells into entirely new sheep, thus guaranteeing that the animals would produce whatever desired protein was needed, rather than relying on the notoriously fickle process of injecting genes into the nuclei of embryos?

The cells Wilmut was after were embryonic stem cells. And while cloning, or nuclear transfer, seemed a logical and even brilliant way of bypassing the inefficiencies of traditional genetic engineering, executing it would require a number of firsts. No one had successfully repeated Gurdon's experiments with mammalian cells, so there was no reason to believe that it would be possible to isolate embryonic sheep stem cells, much less genetically modify them and then clone them using nuclear transfer. ("To this day, there are no embryonic stem cells from sheep," he says emphatically, a reminder, of course, of what shepherds already knew—that sheep are notoriously delicate animals and perhaps not the most ideal species on which to conduct rule-bending studies of development. Sheep cells, it seems, are especially finicky and don't thrive well in a cell culture.)

Wilmut wasn't deterred by these hurdles. In fact, his conviction that marrying nuclear transfer and embryonic stem cell techniques would provide a far more efficient way of genetically modifying animals was only reinforced as he enjoyed the warm glow of a pint and the company of colleagues in a Dublin pub one January evening in 1987.

Attending the annual meeting of the International Embryo Transfer Society in Ireland that winter, Wilmut was learning what was really going on at the lab benches around the world by raising a glass with fellow embryologists and engaging in the off-the-cuff discussions that are often the most valuable part of any scientific gathering. Over the course of the evening, he heard about a scientist in Texas who had more or less successfully cloned a cow. Steen Willadsen, originally from Denmark, had taken a blastocyst cell from a very early, week-old cow embryo, injected it into an egg whose nucleus had been removed, just as Gurdon had done with the frogs, and successfully obtained embryos that appeared to divide and develop normally—at least for a few days—before dying.

For Wilmut, it was as if a bolt of lightning had struck. If Willadsen had successfully cloned a blastocyst-stage cell, Wilmut thought, then it might be possible to take those blastocyst cells, genetically engineer them to ferry a particular protein-making human gene, and then select out only those cells in culture that exhibited a precocious tendency to express that gene. Those cells could then be cloned using nuclear transfer, as Willadsen had done, to generate a live animal guaranteed to be a high-yielding producer of the protein in question. It was a way of turning the haphazard method of creating Tracy into the animal biotech equivalent of a sure thing. Such an animal would be a true clone, rather than a chimera, as Tracy was, since she was the genetically designed product of her sire's egg and sperm and Wilmut's introduction of the AAT gene.

The key to Willadsen's success, it turns out, was stem cells. The remarkably morphing cells he had lured out of the blastocyst possessed a latent ability to become all of the cells in the adult animal, and when transplanted into an enucleated (nucleus-free) cow egg, they had done just that—begun to generate what might have become a new calf.

Wilmut hoped to exploit the same cells, and manipulate them to carry specific genetic changes that would produce pharmaceutically useful proteins, except in a species far more plentiful in his environs—sheep. "We hoped initially these would be embryo stem cells," he says, "so we had a biggish project funded by a mixture of private and government money to try to both derive stem cells from sheep embryos and do nuclear transfer from them."

At the time, however, the science of embryonic stem cells was hardly a robust field with a thick network of researchers to nurture and encourage it. The very concept of a stem cell, in fact, had only recently been born in the labs of a mouse biologist tucked away in Bar Harbor, Maine.

★ ★ ★

In 1953, about same time Gurdon was beginning his breakthrough studies on *Xenopus* nuclear transplantation at Oxford, Leroy Stevens, a developmental biologist, arrived at the Jackson Laboratory along the coast of Maine and was presented with what he later called a "crazy" assignment. His lab leader had received a grant from a major tobacco company to investigate the potential carcinogenic properties of cigarettes—or, more specifically, of the paper that the tobacco was rolled in, which company officials believed might be the true cancer-causing culprit. After exposing mice to the various components in cigarettes, including the tobacco, he stumbled upon a grotesquely fascinating tumor in the scrotum of a commonly used strain of lab mice, strain 129. Scientists had known that male mice in this group tended to form tumors spontaneously, although they didn't know why. And they didn't much care; the cancers were more of a nuisance than a cause for investigation, since the growths essentially contaminated experiments and forced the researchers to start all over again. In most labs, the mice were sacrificed.

But the sheer size of the tumors that Stevens was seeing in a few of his animals piqued his curiosity, so he performed an autopsy and sliced open the gonadal tissue. What he found was a fantastically freakish amalgam of some of the body's various tissue and cell types: traces of muscle, bone, cartilage, skin, teeth, tubules that went nowhere, gut and nerve cells—a biological shop of horrors. "It almost looked as if you'd taken the embryo and chopped it up and rearranged it in crazy ways," says Gail Martin, a professor of anatomy at University of California, San Francisco, who would build on Stevens's work. Stevens also found that in addition to the wild combination of different tissue types, the tumors contained pockets of undifferentiated stem cells, the tabula rasae of the cellular world, which start out as blank slates but through the course of dividing and growing eventually give rise to all of the different cell types in the body.

While Gurdon homed in on the egg and its ability to reverse development, Stevens had tapped another promising vein of regenerative ore: tumor cells. Clearly, the biological soup of tissues in the tumors provided a powerful platform for studying how those stem cells turned themselves into differentiated tissue—or, in the case of cancers, took a molecular wrong turn in development. At the time, there was precious little in the way of developmental maps that tracked a cell's progression from an undifferentiated to a differentiated state.

Conducting a series of experiments in which he methodically sliced off and then injected groupings of the tumor cells into mice, Stevens managed to peel away the stem cells from the junkyard of tissues in the tumor, and isolate the ones that were responsible for morphing into the aberrant array of tissues in the tumor. He called these stem cells embryonal carcinoma (EC) cells for their ability to seed cancers.

But was this biological plasticity a unique feature of the gonadal germ cells, or even of the tumor? Cancer is, after all, a disease in which the cells become stuck in a perpetual "on" position, dividing, growing, and dividing again and again with abandon. And the EC cells came from cells in the testes that had gone malignantly rogue.

Stevens's work suggested that if the tumor he had studied arose from a stem cell that triggered an uncontrolled explosion of growth in a bunch of different tissues, then maybe a similarly undifferentiated mother cell in early development would give rise to all the tissues in a normal embryo. If normal embryos harbored undifferentiated cells, then isolating these would prove a boon for developmental biologists, giving them a cell they could now study in order to understand how the merging of egg and sperm created first one cell and eventually an entire being of millions of cells. But no one had yet answered the obvious question: Could you extract a truly undifferentiated cell directly from an embryo?

It would take Martin Evans, a geneticist at the University of Cambridge in England, and Gail Martin, in San Francisco, nearly a decade to simultaneously supply the answer. Working separately on their respective continents, Evans and Martin (Martin had recently completed her Ph.D.

in Evans's lab) managed to grow up stem cells extracted from a three- to five-and-a-half-day-old mouse embryo on a specially concocted broth of nutrients, cells, and nurturing sera.

In Martin's case, success was a matter of serendipity. Day after day, she mated mice and flushed blastocysts from the pregnant females seventy-six hours after she detected the copulation plug that indicated successful fertilization. She surgically scraped away the inner cell mass of stem cells in the days-old blastocysts. Then she painstakingly extracted the stem cells from these inner cell mass structures with a micropipette, sucking them up one by one and plating them on a culture dish. And day after day, she watched the cells die.

Finally, in frustration, she recalled a trick she had used as a graduate student. While earning her degree, Martin had become adept at working with tumors in another species—chickens—that required quite a bit of coaxing to grow in the lab. They needed a molecular cheerleading squad goading them to divide and grow. "What they needed was another cell type to provide them with factors—and I didn't know what," she says. Eventually, she learned that the chicken tumors simply required some reminders of home, in the form of feeder cells that mimicked their normal avian environment. Irradiated mouse fibroblast cells, which are the most common cells in connective tissue such as the skin, did the trick. The mouse cells could not divide, but they did release nutrients and other growth factors that convinced the chicken cells to divide.

"One day I thought, 'Eh, I'll throw the [mouse] stem cells in fibroblasts to see if they could make the stem cells grow," says Martin. "And lo and behold, as long as I added the fibroblast feeders, I could get individual stem cells to grow and form big colonies. If I didn't add the feeders, they just died."

Martin published a description of the new cell lines in *Proceedings of the National Academy of Sciences* in December 1981. To distinguish her cells, which she grew from stem cells taken from mouse embryos, from the embryonal carcinoma cells that Stevens and others had extracted from the gonadal tumor cells, Martin called hers embryonic stem cells—and is

credited with coining the term that still is associated with the very earliest stem cells isolated from animal or man.

★ ★ ★

Evans's and Martin's papers quickly launched a flood of new studies and
laid the foundation for research into stem cells, but that field had nothing to do with curing human diseases—at least not initially. With the
ability to grow an infinite number of mouse embryonic stem cells indefinitely, scientists set their more immediate sights on answering questions
of embryonic development. It would take more than a decade for the
technology's potential as a source of fresh and healthy cells to replace
damaged or deficient ones in disease to bubble up as a possible advantage.
So while the embryonic stem cells did usher in a revolution in biology, it
just wasn't the one that we are now familiar with.

"People were not interested so much in the fact that these were stem
cells and that maybe they could make tissues from them," says Martin.
"Nobody was really working on it. People were excited about the fact
that you could take embryonic stem cells and put them back in the blastocyst and get a normal mouse. With these cells, it became possible to
mutate genes in these cells—the whole revolution in making mutant
mice by mutating embryonic stem cells and then making animals out of
them that carried these mutations; that was the be-all and end-all."

Martin's and Evans's work made it easier for researchers to study the
effects of genes, by allowing them to knock out specific genes in cells;
generate new, genetically engineered animals from them; and observe
how their manipulations altered the animals. And while embryonic stem
cells are now seen as a potential source of healthy cells to treat disease,
many of these knockout mutations were designed to do just the opposite, actually causing disease in mice in order to study the effects the
changes had.

In a way, Wilmut, too, followed this same path that the earliest
biologists took from Martin's and Evans's remarkable feats. But instead
of knocking out genes from embryonic stem cells and generating new

animals from the resulting genomes, he wanted to add in genes to the stem cell lines and grow sheep, like Tracy, that had been deliberately and precisely engineered to produce certain proteins.

His director at Roslin, however, wasn't as convinced this would work. Several other groups in the United States were already pursuing the same technique with cows, and he thought Wilmut was joining the race too late.

But it was simply too powerful an idea for the determined embryologist to give up. Admitting to a bit of naive optimism, Wilmut managed to convince his skeptical director that the field was undefined enough that anyone entering would have some hope of bringing fame, and maybe even fortune, to their home laboratory.

It would take a year to find a way around the vexing problem of getting sheep embryo stem cells to grow in a culture dish once they were removed from their natural environment. After a few rounds of cell division, they inevitably began to march off in different specialized directions. Try as they might, Wilmut's team could not keep the cells in their embryonic state.

Which is why he and his partner, Keith Campbell, found themselves picking up where Gurdon left off and contemplating whether a fully differentiated cell from an adult animal could do the trick just as well. Campbell, a long-haired free thinker who had begun his scientific career in developmental biology—studying *Xenopus* frogs—was making a full circle back to his roots after detours into pathology while studying in Yemen and fighting Dutch elm disease in Sussex. His laid-back but insightful demeanor was well suited to Wilmut's often meditative distraction and wandering curiosity. And at an introductory lunch at Roslin, Wilmut was struck by Campbell's innovative ideas about making nuclear transfer work—despite other experts' opinions to the contrary—in mammals.

The key, says Wilmut, was Campbell's idea to synchronize the cell cycles of the donor and recipient cells. Cell division is a miraculously choreographed series of myriad steps as each of the cell's chromosomes unspool from its tightly wound ropes inside the nucleus, replicate, then recombine in perfect position in the center of the cell before the cell pinches off, like a fluid hourglass, to divide in two.

Perhaps the success of nuclear transfer, Campbell surmised, was dependent on syncing the phases of the donor cell and the recipient egg, so that both were in a quiescent state until they were united in the lab dish. As in a perfectly coordinated ballet, it wouldn't do for the principals to be dancing to one score while the orchestra was intently playing another. For the egg, the ideal quiet state turned out to be just before it left the ovary, where it conducts most of its development from immature follicle to fully developed egg. For the donor cell, that stage was a resting phase in between cell divisions that Campbell managed to attain by starving the cells into inactivity. Lining up the two cells at these stages should, he thought, increase the chances of their merging and beginning to divide in synchrony, leading to a properly functioning embryo.

Excited by this new approach, but still a bit uncertain about trying it right off with an adult cell as the donor, Wilmut and Campbell decided to take baby steps first. For the donor cell, they turned to one of their short-lived embryonic sheep cells growing up in a culture dish. The cells didn't stay undifferentiated for long, but for this study, a cell that was already on its way to developing into a mature one was perfectly acceptable. The result were Megan and Morag, two Welsh mountain ewes that were born in the summer of 1995. They were the only two lambs born out of 244 embryos that Wilmut and Campbell had created.

Now confident that the timing of cell cycles was the key to making nuclear transfer work in sheep, they convinced PPL Therapeutics to fund the next experiment—the one that would involve cloning a more developed, older cell; the one that would yield Dolly. By this time, the company was just as keen as Wilmut and Campbell to test the radical new idea for optimizing nuclear transfer. It didn't hurt that PPL's director at the time, Alan Colman, had completed his Ph.D. under John Gurdon, and having been trained in the theories of regenerative biology, had a personal interest in and curiosity about the outcome of the Roslin team's nuclear transfer experiments.

In the dead of winter in a tiny room on the ground floor of the Roslin Institute, Wilmut's two micromanipulators—with far steadier hands than he—performed the delicate and tedious process of surgically removing the nucleus from an egg no larger than the tip of a needle and replacing it with the nucleus from another cell. They did this over and over and over—277 times, in fact. That's how many udder cells from a deceased Finn Dorset ewe the team transferred; the mammary cells from the six-year-old sheep had been removed and frozen as part of a study of sheep lactation.

To Wilmut's amazement and disbelief, six weeks after the resulting twenty-nine embryos had been transferred to their respective surrogates to carry to term, one of them, destined to become Dolly, survived.

"It turned out that nuclear transfer is much more powerful than we had anticipated," Wilmut now says with characteristic understatement, "and that we could clone from adults. That definitely was not our specific objective when we started the project. We would have seemed outlandish and silly at the time if we had proposed it."

But scientific breakthroughs often emerge from what appears on the surface to be ridiculous, and as Wilmut's team nervously monitored the surrogate's angst-inducing 153-day pregnancy, the impossible—cloning a mammal—became possible on July 5, 1996.

★ ★ ★

Dolly lived an eventful life of six years before passing away on Valentine's Day in 2003. Wilmut says it was he who had to make the difficult decision to end her life. She was suffering from a nasty case of pulmonary adenomatosis, a viral infection that caused tumors to blossom in her lungs; Wilmut could not watch his special charge suffer any longer.

But even in death the pioneering ewe continues to raise provocative questions. Most sheep live for more than a decade; Dolly made it only to six years. How old *was* Dolly? Was she as old as the udder cell from which she had sprung, plus the six years she had lived as a clone? Would her cells reveal that she had lived not one, but two lives? An autopsy found that she had suffered from arthritis—more a reflection of her hefty weight, the result of her penchant for procuring treats from her dozens of admirers, says Wilmut, than a consequence of any fast-forwarded aging. On a molecular level, however, her cells did show the classic signs of advanced age: The telomeres, or tips of DNA on the ends of her chromosomes, were shortened, just as they are in older sheep. But she hadn't died of old age; she'd died because she'd been unfortunate enough to catch a bug that was circulating in the barn where she'd lived.

In the years since Dolly's birth, a stunning array of other mammals have been cloned in the same way—the list includes mice, cats, cows, pigs, goats, horses, and dogs—but it's still not clear exactly how or even why these clones actually exist. "In a way, we should still be surprised that nuclear transfer works rather than disappointed about the low efficiency that it has," notes Wilmut.

Fundamental questions about how faithfully the cloning process recreates a viable embryo still remain, some of which may have implications for the usefulness of nuclear transfer as a source of stem cells to treat human disease. Researchers have documented a condition called "large offspring syndrome" in almost all cloned animals, for example, in which the newborns are significantly heavier than their normally conceived counterparts and gestate with remarkably large placentas that often cause miscarriage for the surrogates. Their lung development is also often abnormal, causing them to pant and develop respiratory distress that can lead to early death.

But such differences emerge only when nuclear transfer is used to clone new animals. What about the stem cells that could be extracted from a cloned embryo? This is what Wilmut and Campbell were after when they began their experiments—not a full-blown animal clone but something perhaps a little less biologically ambitious, a long-lived, reliably rich source of pluripotent cells that they could manipulate for genetic engineering.

But even at this cellular level, differences in the stem cells coming from nuclear transfer embryos were beginning to emerge that distinguished them from stem cells derived from IVF embryos. As cells become increasingly differentiated, their entire genetic code takes on a unique molecular pattern, not unlike the handwritten notations that a musician might add to a musical score as reminders to play more softly or add more depth when executing certain passages to improve the expression and interpretation of the melody. Called methylation, the process involves depositing a chemical compound known as a methyl group, composed of a carbon atom and its hydrogen partners, into the chemical base pairs that make up DNA. Methyl groups mark up a cell's genome with a specific, unique interpretation of the genetic code that allows it to follow a certain lineage toward becoming more specialized. Like those handwritten reminders scratched in by the musician, such methylation patterns do not permanently change the genome of a cell the same way that a mutation does, but they can be passed on to daughter cells when a cell divides.

Methylation, it turns out, is only one example of these so-called epigenetic changes—alterations that don't occur to the genome itself but that affect how the genes are expressed, or activated. Similar to the way the wrinkles and freckles that people accumulate are signs of their experiences throughout life, epigenetic changes reflect the environmental exposures and other legacies of a cell's journey from the embryo onward. These changes can affect the way DNA assembles itself into chromatin (the ropelike structures that package the genes inside cells), as well as the length of the telomeres.

Reprogramming, as accomplished by nuclear transfer, is supposed to wipe all of this epigenetic legacy of an adult cell clean, bringing it back to its pristine embryonic state. However, researchers such as Rudolph Jaenisch, a biologist at Massachusetts Institute of Technology and a pioneer in understanding the molecular differences among stem cells, is not convinced that the reprogramming faithfully erases all of the methylation that an adult cell accumulates. In fact, he presumes that's why the 276 other embryos Wilmut transferred didn't make it beyond the initial first steps in the process—the embryonic stem cells failed to recapitulate all of the epigenetic changes that each of the adult sheep's various cells had accumulated in order to become their specialized selves.

But Jaenisch is also not convinced that strictly re-creating the epigenetic legacy of a cell is necessary if your goal is not to generate an entirely new animal but to amass a useful culture of embryonic stem cells instead. To grow a line of stem cells, it may not matter if the cloning process can generate all of those cells faithfully; all you really need is a set of cells that can remain in a pluripotent state long enough for you to then coax them into becoming whatever specific cell type you want to study—neurons or muscle cells or heart cells.

Whether or not Jaenisch is right is a weighty question that stem cell scientists continue to ponder, even as they push forward to generate as many different types of body cells as they can from these human embryonic progenitors. Even without an answer, they say, the cells could still be a rich source of previously inaccessible information about how normal cells mature and what happens when that process is tainted by disease.

That's something that Wilmut himself had in mind when he embarked on his cell-transferring efforts, even if this potential was overshadowed by the public's excitement over the more sensational idea of cloning. "Although we played it down at the time when we announced Dolly, we were already thinking that cloning to get human embryonic stem cells could be a new way to study disease," Wilmut tells me. That was a hard message to convey, however, when every citizen and every legislator could see little about Dolly's birth beyond the perils of cloning. Dolly wasn't simply a target of scientific fascination and amazement, but a bull's-eye for political and moral tensions as well.

3

Politics Intrudes

Dolly's birth set off a frenzy of animated discussion—at dinner tables, in coffee shops, at universities, and in legislatures—about the inevitable question of what would come next. After the initial flurry of media attention, Dolly herself almost became an afterthought, as what really set people's minds racing was the idea of a human version of the history-making ewe. Cloned farm animals were one thing, but a cloned human baby—that was quite another. President Clinton quickly diverted the National Bioethics Advisory Commission, which was busy discussing the protection of human subjects participating in research trials and the appropriate use of genetic information, to consider the implications of human cloning. Giving the commission ninety days to come up with recommendations, Clinton in the meantime addressed public unease over the specter of human clones by imposing a moratorium on federal funding of all such research, and implored private interests to voluntarily adhere to the same ban on pursuing such experiments.

On February 26, 1997, the Wednesday following the public announcement of Dolly's birth in the journal *Nature*, Harold Varmus, director of the National Institutes of Health (NIH), found himself addressing Congress in one of NIH's scheduled appropriations hearings to determine its annual budget. The timing could not have been better for the popular director to address the news of Wilmut's remarkable achievement, which was buzzing all over the broadcast and print media. It was an opportunity, Varmus realized, to nudge the public conversation about cloning beyond the prospect of generating biological copies and instead frame the discussion around something far more promising: stem cells. And he wasn't going to miss it.

How valid, the elected representatives wanted to know, was Wilmut's feat, and how soon could his technique be used to clone a human being?

Dolly clearly made Congress nervous—cloning was a science that seemed to cry out for some type of guidance and regulation, along the same lines as the use of human subjects in medical research and the advent of DNA techniques enabling man to manipulate and refashion DNA into never before seen combinations. But what sort of guidance and regulation? How does a government, as a representative body of a populace, set standards of practice and boundaries on what is and is not acceptable to do with such a powerful, unprecedented technology? How does one regulate the creation of human clones? Should the government be funding such research? Or even more to the point, should the government, representing the views of the people, ban the science altogether as too dangerous, too irresponsible, and too damaging to our entrenched views of human dignity, human autonomy, and the sanctity of human life?

As director of the NIH, Varmus was responsible for dispensing more than twelve billion dollars toward basic biomedical research, the largest investment behind such experiments in the country. It would rest on Varmus to guide the NIH, and thus the government, through the coming white waters of confusion and speculation that was sure to surround the promise and potential of cloning. As nervous as biologists were about the growing public clamor over what Wilmut had done, they were confident that if any scientist could defend the research before Congress, Varmus could. A popular choice when he was appointed in 1993, the tall and wiry geneticist had earned their respect because, like them, he had logged long hours in the lab and appreciated the value of basic research. Still fit at fifty-seven, he retained the physique of a high school athlete, and indeed was well known for donning his racing helmet and cycling gear to bike twelve miles to and from work in Bethesda, Maryland, every day.

Like any adventurer setting out for uncharted territory, Varmus found himself drawing a crude map at that hearing of the truly brave new world of biology that Dolly's birth now made possible. Cloning, he explained in a lengthy and energetic lecture that continued for several pages in the hearing transcript, would be only the first step in venturing forth on two very different but equally exciting paths. Most obviously, the ability to replicate an entire mammal would become a powerful tool in many fields, not the least of which was agriculture. Wilmut's efforts had begun, after all, with PPL Therapeutics' desire to find a way to turn sheep's milk into medical bounty, by creating ewes who were genetically designed to produce milk laced with almost ready-made pharmaceutical products. And while plant biologists had been cloning species for decades to generate hybrids more resistant to drought, infection, and all manner of punishing conditions, the nuclear transfer technique that Briggs and King had first stumbled upon, which Gurdon had refined in amphibians and Wilmut had now brought to mammals, would make such optimization of desired features a reality for mammals, as well. Rather than relying on the crude and unpredictable practice of inserting and deleting genes from a genome, nuclear transfer would now make breeding the choicest meat-producing cattle and the richest and highest-producing milk cows a far more reliable endeavor.

Even beyond cloning entire animals, however, was the second enticing journey that Varmus could only begin to sketch out for the members of Congress at that hearing—the one that, to his mind, was more intriguing. This route promised a tantalizing new vista of biological possibility, not only with agriculture but also with human cells, although at the moment it seemed more fantasy than possibility. If a fully mature mammary cell could be nudged back into an embryonic state in the ultimate show of developmental reverse engineering, then all of a sudden reprogramming began to take on a whole new meaning. The entire foundation of how we thought about and tackled human disease could also be rebuilt in a profoundly significant way. No longer would therapies have to rely on addressing symptoms or merely cleaning up in the aftermath of chronic conditions such as diabetes or Parkinson's or cancer. If cells could be reprogrammed, then they could also be reborn—and most important, reborn in a healthy, functioning form to replace ailing and deficient predecessors.

This is what Varmus focused on during his testimony—the clinical ramifications of what another witness at the hearing would dub "therapeutic cloning." Therapeutic cloning involves the use of nuclear transfer to clone not for reproductive purposes, but solely for the intent of generating and extracting the embryonic stem cells produced by the resulting embryo. Varmus sensed correctly that it was crucial at this early stage to separate the two uses of cloning in the public's mind. That was the only way he could see to insulate the true promise of cloning from the ethical and political storm that would inevitably consume discussions about Wilmut's achievement.

His conviction came from experience. Varmus had already witnessed firsthand how politics could intervene and redirect science, and he appreciated, perhaps more than even the researchers in the field at the time, just how vulnerable anything having to do with embryos would be to getting pulled into the quicksand of the abortion debates. Ever since *Roe v. Wade* had been decided in 1973, federally funded research on fetal tissue or human embryos had been held hostage by a deadlocked legislature. While liberals favored allowing taxpayer dollars to support studies on fetal tissue and human embryos that could improve our understanding of disease and benefit reproductive science, conservative legislators had been working tirelessly to build a base of opposing legislation and portray the Supreme Court decision as an anomaly that would one day be overturned.

Within weeks of Varmus's arrival at the NIH, in fact, President Clinton lifted a five-year ban on the use of federal funding to transplant human fetal tissue. The moratorium had been put in place by President Reagan in 1988, and reinstated by President George H. W. Bush. It effectively prohibited any substantive studies on fetal cells or embryos to treat disease, because the primary source of these tissues was abortions.

Clinton wrote his decision into law as part of the National Institutes of Health Revitalization Act, and the NIH was now free to fund fetal and human embryo research again. So one of Varmus's first tasks as director was to convene a panel to advise the institutes about the best way to allocate taxpayer dollars for studying human embryos in a legal, safe, and morally acceptable way. While its members had no way of anticipating the extraordinary breakthroughs that would come, one after the other, in stem cell research over the next five years, the panel's decisions would leave a lasting and profound footprint on the regulatory and legislative landscape of the emerging field. The panel's conclusions eventually formed the core ethical guidelines on stem cell research that organizations such as the National Academy of Sciences, professional stem cell societies, and even private biotech companies would consider when coming up with their own set of principles for conducting research in the field. And the disappointing reception that the panel's final report received in the Clinton administration would serve as a prophetic warning about just how difficult it would be to pry the promise of science from the rhetoric of politics whenever embryos were involved. That was the lesson Varmus learned in a very personal and powerful way when his convictions and integrity as a scientist were put to the ultimate test the day he presented the panel's recommendations to the White House.

★ ★ ★

It was a typically chilly February day in 1994 in Washington when a group of nineteen specially selected individuals made their way to the

Marriott on Pooks Hill Road in Bethesda, Maryland. The setting was far from auspicious—a routine banquet hall that doubled as conference space in a standard suburban hotel—but what those nearly two dozen experts were there to do was far from ordinary. Among the bundled-up group arriving that frigid morning were a handful of physicians, a few ethicists, a couple of lawyers, some political scientists, and some patient advocates. Right away, each of them knew that there was something different about this panel; the cluster of journalists, NIH officials, government personnel, and other interested citizens who greeted them was a giveaway. Their presence meant that the discussions and inevitable verbal tugs-of-war that were the lifeblood of such gatherings would be chronicled and scrutinized in the press as well as in the halls of the legislature and the White House. The attention was far from standard fare for a matter of science. But this matter involved embryos, and that made all the difference.

Each of the panelists had been chosen by Harold Varmus to confront the first major challenge of his tenure as director of the NIH. He wanted, he told them at that first meeting, to answer a series of important what-if questions.

By this time, the first test tube baby, Louise Brown, born after British scientists united her father's sperm with her mother's egg in a petri dish, was a teenager, and the process of in vitro fertilization that had created her had blossomed into a billion-dollar business in the United States, with more than 350 IVF clinics and tens of thousands of babies following in Brown's test tube footsteps. Creating embryos outside the womb was now standard practice, and scientists were not blind to the fact that in addition to being a welcome solution to infertility, these embryos provided a potentially powerful new source of information about human development. And that understanding extended beyond simply improving upon the basic reproductive process.

What IVF made possible was research on an embryo even before it was implanted in a woman's womb, something that had been up to this point completely impossible to do. Varmus wanted his blue-ribbon panel to consider, and recommend to him, what to do with these human embryos that had been created outside the womb.

What if, for example, eggs could be activated, as they had been in some cases in animals, to divide and become an embryo without first being fertilized by sperm? What if, at some future date, it became possible to create embryos by some sort of cloning process, in which an adult human cell inserted into some specially manipulated egg could result in a viable genetic copy of the original donor? And what if, in order to understand human development or even disease processes further, scientists wanted to generate embryos from women with, say, diabetes, who donated their eggs, and men with the same disease who donated sperm, with no intention of creating a baby but instead to provide researchers with a way to study diabetes as it developed, from infancy on? What if all of these scenarios were possible?

Now that the NIH had received the green light to fund embryo research, Varmus wanted to know how the government, as gatekeepers of the largest flow of research dollars in the country, should regulate such activity. And how would the American public, as the investors in these studies with their tax dollars, feel about the direction these experiments would take—culturally, ethically, and morally? Would such manipulation of the basic materials of human life be acceptable? Were they simply the inevitable extension of the march of science, the evolution of in vitro fertilization into far more than an answer to infertility? Or was there a bright moral line that society would or should draw, beyond which scientists dared not tread?

These were the heavy questions that Varmus asked the panelists to consider, and he didn't expect easy answers. The fact that the group was meeting at all was a victory of sorts. The lengthy shadow of abortion was a political reality that the nineteen experts sitting around the conference room table would be reminded of, painfully, over the coming months. It was a reality they could not escape, and it would be no different when the first human stem cells were isolated, from an embryo, four years later. Even before IVF or cloning or stem cells entered the medical lexicon, the political debate over abortion was already defining the science of human embryo research.

* * *

Following Louise Brown's birth in 1978, when it became clear how monumentally significant the new technology of in vitro fertilization would become, Congress quickly convened a federal ethics advisory board (EAB) to address the uncomfortable questions about deliberately matching up egg and sperm in the lab. Its job was to consider whether the U.S. government should fund research in the area of in vitro fertilization, and if so, what type of studies would be acceptable. With the baby, as it were, already out of the test tube, and the unavoidable emergence of assisted reproductive technologies as a viable medical platform, the NIH needed to know how Americans wanted their research dollars spent. How much study on the earliest forms of life would the public be willing to support? What types of research were acceptable? How far could scientists go in creating and manipulating embryos?

By 1994, however, the federal government had yet to truly confront such ethical questions or develop a policy to guide research on embryonic tissue. Since 1974, federal health officials had brushed the issue aside after imposing a de facto moratorium on IVF research, awaiting the EAB's recommendations for how to proceed. But while policy makers chose to coast along with this regulatory limbo, thousands of infertile couples did not. Pumping cash into a fledgling industry, desperate couples routinely spent anywhere from eight thousand to twenty thousand dollars for a full cycle of treatments that nationally yielded only a 15 percent or so chance of bringing home a baby. The upshot was that even without government regulation or financial support, the IVF field had blossomed entirely in the private sector, in the labs and clinics funded by patients and business investors eyeing a burgeoning new market.

More than a decade later, the still unresolved political debate over using human embryos for research would push stem cell science along a similar path, out of the government's purview and into the unregulated and commercially driven private sector. It wasn't obvious then, but the same political tension about embryos would reemerge, in a frustrating repeat of history, when scientists isolated the first stem cells from a human embryo. The precedent had been set, at least at the federal level, to keep studies of the earliest stages of human development at arm's length.

It began when the EAB came back with its report after several months of discussion in May 1979. The board ended up supporting a wide range of study on human embryos, including the creation of embryos purely for research purposes, in which there was no intention of ever implanting the embryo into a woman to achieve a pregnancy. There was one condition, however. Before any federal funds were granted, each experiment would have to be reviewed and approved by the EAB. That would be in addition to the local review that academic institutions generally require of experiments that occur under their auspices—an evaluation that includes a thorough vetting of the scientific rigor of the study, as well as assurance that subjects had been fully apprised and were aware of the risks and benefits of participating in the investigation.

The EAB review, while both reasonable and responsible, would nonetheless prove too contentious politically. No sooner did the report become public than the uproar began. Nearly thirteen thousand letters streamed into the Department of Health and Human Services, then directed by Joseph Califano Jr., castigating the agency for even considering spending taxpayer dollars on IVF research. Citing violations of moral and predominantly religious principles honoring the sanctity of human life, these missives made it clear that IVF research, at least under government auspices, faced an uphill battle against a very vocal conservative contingent of the population.

Snowed under by the volume and vehemence of the opposition, Califano decided to leave things as they stood. Fighting to keep his job in the cabinet (he had alienated some of President Jimmy Carter's strongest supporters with his aggressive positions against smoking and in support of desegregation), Califano fell victim to a stunning cabinet purge as Carter reshaped his administration for a tough reelection campaign that fall. Two months after the report was issued, Califano was fired from his post. When the EAB's charter expired at the end of the year, it was not renewed. Without funding to keep the board going, in 1980, after just a brief run, it ceased to exist.

All of which led to another de facto moratorium on federal funding of IVF and related embryo research. Without the EAB's approval, no studies could be approved, and without an EAB, no studies could even get reviewed. And after Carter's defeat in November 1980 by Ronald Reagan, a strong opponent of abortion, the possibility of further studies on fetal tissue, and by association human embryos, was pushed aside yet again. In 1988, affirming his opposition to the use of any tissue from aborted fetuses for research, Reagan imposed a formal moratorium on federal grants in the field.

The most cutting-edge research on the earliest stages of development, then, slipped out of the public's purview and into the hands of the private sector—businesses such as IVF clinics that by definition were driven more by commercial interests and financial incentives than scientific pursuits. "I always thought that was a real shortcoming of American medicine," Varmus tells me over breakfast one fall day near Memorial Sloan-Kettering Cancer Center, where he served until 2010 as president between his tenure at NIH and his return to Bethesda to head up the National Cancer Institute. "For years, NIH was unable to support any work in IVF. We weren't doing as well as the Brits in an indisputably important field."

Decades later, that history would repeat itself when the same political reticence to embrace embryo research meant that the United States would find itself playing catch-up to other nations, including the United Kingdom, when it came to devising a consistent and coherent policy of stem cell research. And while the breakthrough in isolating the first human embryonic stem cells occurred in a university lab, it was financed with private funding from a biotech firm, Geron, and philanthropic donors.

★ ★ ★

In 1993, with Bill Clinton's election to the White House, however, neither Varmus nor the members of the panel he convened at the Bethesda Marriott could have anticipated that unfortunate scenario. With a change in administration, the political winds had shifted once again, this time in favor of science, and the medical research community, Varmus included, were optimistic. Even the name of the appropriations legislation was an encouraging sign that underscored exactly what the influx of dollars was intended to do—reinvigorate the agency into funding innovative science. The law gave scientists fresh reason to push for previously off-limits areas of research—in HIV, in reproductive health, on embryos and fetal tissue. Most important for embryo and stem cell research, as part of the Act Congress eliminated the EAB requirement, thus unlocking the restriction on human embryo research and giving NIH the green light to begin funding these studies again.

Which is why Varmus needed his panel. He needed some type of regulatory map to navigate the ethically contentious waters of experimenting at the earliest stages of human development.

That February morning, he instructed the members to consider embryo research in all its different forms—in vitro fertilization, spermless embryo formation, and the like—and to advise him on three categories of experiments: those studies on human embryos that the NIH should fund; those the agency should not support; and finally, experiments that the panel believed were worth considering but warranted further review by another body before NIH dollars were invested in them.

The panelists were not asked to consider stem cell research per se, but in very short order, their discussions expanded to include the potential of these cells. Without intending to, the Human Embryo Research Panel (HERP) ended up becoming one of the first to seriously consider the ethical and regulatory challenges of working with these cells. And it also ended up laying the groundwork for regulating stem cell research that future bodies, both public and private, would come to rely upon for guidance.

*   *   *

The majority of the members, particularly the nonscientists in the group, would hear about stem cells for the first time during their discussions over the ensuing four months. That was certainly the case for R. Alta Charo, the Wisconsin law professor who had never met Varmus but was aware of his scientific credentials and eager to get started on what she considered groundbreaking work in mapping out policy for human embryo research.

Compact and efficient, with a head of short, wavy dark hair, Charo is a native New Yorker who talks fast and thinks even faster. Well respected among leading legal and policy experts, she would later be among the first people the Obama administration consulted when the newly elected president was formulating his own stem cell policy. She would also be one of the first legal and ethical scholars James Thomson sought in 1998 to discuss the social implications of his breakthrough work in isolating the first embryonic stem cells from a human embryo.

But in 1994, Charo had yet to wade into this world and came to the panel with a strong background in women's reproductive health issues. Having logged time at the Congressional Office of Technology Assessment, where she focused on legal issues surrounding the impact of reproductive technologies on the family such as surrogate motherhood and gamete donation in in vitro fertilization, she thought she would be considering embryo research primarily in the context of reproductive health.

As a lawyer, Charo was intrigued by the complex legal relationships growing out of the new technologies, and the implications that they would have on legal rights and concepts of family and family relationships. Once women donated their eggs and men donated their sperm, how much legal right did they have over the fate of the resulting embryos? How far should such legal rights extend? And as a lawyer who had participated in the debates around crafting of policy for contraception and abortion rights not just in the United States but also around the world, she was viewing the emerging studies on embryos through the lens of the evolving feminist movement.

To her delight, however, it became clear that the potential of embryo research would extend far beyond reproductive and women's rights issues. "This is where I first began to hear about the prospects of embryonic stem cell research," Charo says during a thoughtful but typically fast-paced conversation. Despite the panel's priority in viewing embryo research through the prism of improving infertility treatments or even creating new contraceptive options, over the course of their monthly sessions and follow up e-mails it became obvious that studying embryos was only a gateway to a new world. And that world was just being infiltrated by talk of stem cells, whose potential in mouse studies was already leading some scientists to new ways of thinking about and treating nearly every human disease, from neurological disorders such as Parkinson's to the most common chronic diseases of the heart. One of these experts sat at the very same table with Charo and the HERP members.

Brigid Hogan, a mouse embryologist at Vanderbilt University at the time, was forever changed by Martin Evans's and Gail Martin's milestones in picking out embryonic stem cells from mouse embryos. James Thomson, at the University of Wisconsin, was still a year away from making his breakthrough in doing the same in primates, and several years out from similar success with human cells, but scientists like Hogan were convinced those milestones were not just inevitable but also not too far off in the future.

Like Varmus, she was fascinated by the potential of these earliest delegates of development, and serving as the panel's cochair, she couldn't help but infect her colleagues with her excitement and ensure that they looked not just at the reproductive potential inherent in embryo research, but beyond it to encompass, at least theoretically, what she saw as its true value: its ability to impact virtually all human diseases, from heart failure to diabetes to Parkinson's and even cancer.

Hogan had reason for her expansive optimism. So enthralled with the mouse embryo that she carries an image of one in her wallet ("to look at if I'm feeling low," she says), she used mouse embryonic stem cells to discover key drivers in the early stages of growth in embryos. She knew that it would simply be a matter of time before similar techniques could be translated to understanding human development.

Often taking pen to paper—or, more accurately, marker to poster board—Hogan would diagram the genesis of embryonic stem cells for the panel members: how they arose in the days-old blastocyst, and how, at least theoretically, if someone managed to isolate them from a human embryo, they could be cultured in a dish and treated with the right growth factors, enzymes, and other chemicals to begin dividing and developing into nerve or pancreas or liver or muscle cells. She also introduced the thrilling possibilities of nuclear transfer, or what would, in just two short years, with the birth of Dolly the sheep, become known as cloning. At the time of the panel's discussions throughout the winter and spring of 1994, the technique had worked only in frogs, but, Hogan told her colleagues in her cheery British clip, there wasn't any reason why it wouldn't one day be possible to do the same in mammals, in nonhuman primates, and eventually in humans. The purpose, she stressed, at least from a scientific point of view, would not be to clone people per se, but to provide a source of stem cells, which could be removed when the blastocyst was four or five days old. If the stem cells came from an individual suffering from a disease, then these could be studied to expose what had gone awry in that individual's cells—the pancreatic cells of a diabetic, for instance, or the brain neurons of someone with Parkinson's. Once they revealed the defect, these cells could then be repaired or replaced with healthy versions to cure the disease.

"There wasn't any deliberate planning on my part," Hogan tells me when I ask her about the genesis of these impromptu biology lessons during the sessions. "It just emerged that many people on the panel needed to understand the basic science, not just of stem cells but about the whole process of embryonic development and things like cloning. They were spontaneous."

For Charo, these lectures were nothing if not eye-opening. As a nonscientist who had primarily considered reproductive technologies from the perspective of legal rights, ownership, and regulatory challenges, those marked-up illustrations were her first glimpse at the potential breadth of embryo research and the range of studies that embryos could support, particularly through stem cells. It was immediately obvious how valuable studies on human embryos could become, and not just for improving reproductive science.

"All the presentations [we heard] were in terms of the mouse model," she says of the state of the science at the time, demonstrating a very quick grasp of the limitations of the studies. "And while mouse embryos are very valuable in this research, their morphology is different enough from humans that there is a certain point at which you can't learn what you want to learn in the mouse model."

Moving from mice to men, however, was proving problematic, not just on the scientific front but socially and politically as well. "We were trying to figure out the state of the research, and how important it is to do research on human embryos, and how important it is to do it at this time," she says, "because we understood that the research was highly controversial since it was obviously tied to the debates around abortion."

Thomas Murray, a bioethicist who was then at Case Western Reserve University, still recalls how charged the subject of embryo research was, and how tense matters became with the public, many of whom were opposed to abortion, attended the meetings, and observed every exchange. At the time, several abortion clinics around the country had been bombed, and physicians performing the services had become the target of violent attacks. The panel members were not spared either: Charo received death threats, and the NIH posted guards at the meetings in case of an outburst. Keenly aware of how emotional the issue had become for some, Murray admits to succumbing to a bit of paranoia when he noticed a slightly disheveled middle-aged man sitting just behind the front row of spectators at one session that spring, close to the conference table where the panelists sat. The gentleman nervously shuffled and reshuffled a plastic bag he had brought with him, and Murray jumped to nervous conclusions. "I thought, 'What has he got in that bag?' We had no metal screening, and anybody could have walked in with anything if they kept it from sight. I thought, 'What if he's got a gun in that bag? What are we going to do?'" The man turned out to be a citizen there to provide testimony, but the experience reminded Murray of what was at stake.

As the members continued to draw and redraw the lines between what types of research on embryos would be acceptable and unacceptable using federal funds, and which types fell somewhere in between, it became clear that they needed a compass, some moral axis around which they could orient the various things that scientists were proposing to do with embryos. The most obvious such hub would be the moral status of the embryo. If the embryo were considered the equivalent of a living human being, then that would afford it certain rights and dignities, despite its inability to assert them, that would render all of embryo research unacceptable. If, however, as the majority of the panel felt, moral status is a dynamic condition as opposed to a fixed feature, in which rights and status are layered on in accordance with advancing biological development, then it could be argued that the panel needed to find some scale along which these rights were accrued.

As the discussions continued, however, it occurred to Charo that the panel's attempt to assemble this sliding scale was akin to Alice's making sense of Wonderland. What feature defined the exact moment when a mass of cells gained enough "beingness" to earn additional rights and status? Was it the formation of the primitive streak, at around fourteen days, which defined the three basic germ layers of human tissue and thus designated a biological marker for personhood? That was what a panel debating similar thorny issues in England a decade earlier had decided, determining that it was acceptable for scientists to study embryos but only until the embryos were fourteen days old.

That panel, led by Oxford philosopher Dame Mary Warnock, contained more doctors, lawyers, and scientists than ethicists and theologians, however, and rather than tackling the stickier task of setting moral criterion for what type of research was acceptable, it took a more practical approach, relying on objective biological criteria, including the primitive streak to guide them. In the UK, that proved effective in helping to establish a straightforward embryo research policy inspired by the Human Fertilisation and Embryology Act in 1990, which later applied to stem cell research as well.

The Warnock report and its recommendations were certainly part of the mountain of materials the HERP members reviewed. For Charo, however, it seemed that the moral issue couldn't be so easily subsumed by biology. At least it didn't seem likely in the United States, where the abortion debates continued to simmer in the political arena and pro-life interests retained enough clout to prevent panels like HERP from side-stepping the question of what, exactly, the moral status of the embryo is.

But, she admits that even today, "I don't see a way around this, except to acknowledge that we can't answer this question and then ask, 'What do we do with that? Where do we proceed from the fact that we can't answer it, and we can't all agree?'"

The impasse reminded her of Lewis Carroll's poem "The Hunting of the Snark," which chronicles the misguided efforts of a hapless group of shipmates in their hunt for a creature that does not exist. "The snark, in my mind, was the moral status of the embryo," she tells me. "One of the things I concluded at the end of the [HERP] experience was that there are some issues on which we are not going to achieve fundamental consensus. We will not reach consensus on the moral status of the embryo, or on the moral status of somebody whose lower brain stem stopped working or is in the post-vegetative state like Terry Schiavo, because we are all driven by an entirely different analysis of what generates moral status."

That analysis, the panel acknowledged in their review, arises from a complex combination of religious, cultural, and social beliefs that is hard to modify or alter, even in the face of contrary objective evidence. The Jewish and Islamic religions, for example, adopt a more pluralistic approach, and while recognizing that a developing fetus should be accorded some respect as a being, believe that it does not earn personhood until months after conception. In Jewish theology this occurs after forty days; in Islam it takes until the fourth month of pregnancy. The Catholic tradition, on the other hand, adheres to a single-criterion approach, holding steadfastly that all life, from the moment of conception, has equal moral status, despite the scientific fact that conception is not actually a single discrete event, but rather a process of sperm fertilizing an egg.

Charo was urging the panel to put aside the issue of trying to determine an embryo's moral status and address instead, from a civic perspective, what policy concerning research on embryos would be acceptable to a population practicing diverse moral, ethical, and religious traditions.

"Rather than have the government adopt one analysis or another, which essentially puts the government on one side or another of what is fundamentally an ethical, philosophical, or theological debate, why not say we can't agree, and we'll never agree?" she argues. "Now, we can ask a slightly different question if we are asking not about moral ethics but political ethics. What is the ethical thing to do to solve a political impasse? In other words, it's more about political philosophy than moral philosophy. Maybe we start the analysis that asks who is hurt and who benefits if we adopt a particular policy, and if we take a look at how benefits and burdens are distributed."

If the issue of how to treat an embryo can't be resolved on moral grounds by consensus, then, Charo believes, it should be addressed from the perspective of public policy—finding the most equitable distribution of the advantages and disadvantages of something such as stem cell research. There are those who support the field and believe in its potential to cure dozens of diseases and relieve millions of patients from pain and suffering, while at the same time, critics of the science are equally opposed to the destruction of embryos involved.

This revelation led Charo to push, by her own admission, rather aggressively for recommending that the government consider funding the creation of embryos solely for research purposes. This would mean that women and men would donate egg and sperm not with the intent of creating a baby, but only to create an embryo that would then be destroyed at just under a week old when scientists extracted its stem cells. Given the direction and promise of the science of stem cells, Charo felt this was a reasonable and logical recommendation for the panel to make, and for the government to follow. And given the position that the panel was taking on the question of the moral status of the embryo, it also made sense to include this option as one that warranted further review. "As a matter of principle, since this body was viewing moral status of the embryo as a function of its state of development, there really should be no difference morally speaking in the status of an embryo left over from IVF and five days old, or an embryo that was deliberately made for research and is five days old," she explains. "If the moral state depends on the stage of development, then there is no difference. On that basis, one can conclude that yes, you can make them occasionally for research."

★ ★ ★

In the end, the nineteen members voted to place the creation of embryos solely for research purposes into the category of "warrants further review"—but only by a slim margin. They agreed that federal funding of such research would have to fall into two very strict categories: Either the studies could not be conducted without generating the embryos, or their creation was necessary for experiments of "potentially outstanding scientific and therapeutic value."

Two of the members, including cochair Patricia King, a law professor at Georgetown University, and Carol Tauer, a professor of philosophy at St. Catherine University in Minnesota, disagreed with the recommendation and wrote dissenting opinions for the final report.

"I felt that to create embryos for research, purely for stem cell research, did not come under the conditions we had already outlined," Tauer tells me. "That was my objection; I thought we had laid down principles that we had agreed on, and here we seemed to be going somewhat beyond that."

Tauer continues to feel that the panel didn't take into consideration the thousands of embryos that exist in IVF clinics, in frozen storage, that could be used for research purposes if their donors decided not to have them implanted. If the panel's own criteria were followed, she says, those should be the first resource for study, and if they did not suffice for whatever reason, then, and only then, could generating new embryos for research be justified. "As far as I knew, there were a lot of alternatives, and it wasn't necessary to create embryos to study," she says.

King, who penned the other dissent, incorporated a sociocultural consideration into her opposition. During the panel's four months of discussions, King was the first to question whether the creation of research embryos would be pushing the American public's sensibilities on these matters too far. Creating embryos for research purposes created serious ethical anxieties that we as a society had not yet addressed and, she feared, would give the public the wrong impression about the intent of the studies. "The fertilization of human oocytes for research purposes is unnerving because human life is being created solely for human *use*," she wrote in her opinion. "I do not believe that this society has developed the conceptual frameworks necessary to guide us down this slope."

King's concerns, it turned out, proved prophetic. When the final report was released on September 27, 1994, the panel's recommendation that NIH fund the creation of embryos for research became a lightning rod for the national media, as well as for other ethicists. The *Washington Post* published an editorial excoriating the panel for its support of research embryos as "flat wrong," "unconscionable," and science that "the government has no business funding."

But even that was just a prelude to the rude awakening that the entire panel, including Varmus, would get when the report completed its bureaucratic journey through the review process at the NIH and finally hit the president's desk a few months later.

By that time, the Clinton administration would be, as Varmus says, "nervous" about a number of things—the fallout from the congressional election, in which Newt Gingrich had succeeded in reestablishing a Republican majority in the House, and the Democrats' slipping hold on voters. A radical idea like allowing scientists to create human embryos solely for the purpose of conducting research on them might simply be too much for voters to accept.

"I knew not everybody was going to like it," Varmus tells me of his reaction after reading the report. He fully endorsed the panel's findings, and appreciated the thoughtful conclusions they had come to. "It's pretty amazing," he says, still marveling today over the scope and depth of the report. "It addressed a lot of issues before they arose. It wasn't just that they thought about stem cells, but that they thought about nuclear transfer, at a time when nuclear transfer wasn't even possible in humans. So I really didn't expect to have this much push-back from our own side."

But pushed back he was—by an administration that he had assumed would be behind the panel's recommendations. After reviewing the report, Varmus trekked several times from NIH's Bethesda campus, along Rockville Pike to Pennsylvania Avenue. "I spent time at the White House, not talking to Clinton, but to people around him, to explain what this research was, trying to present it in as objective a way as I possibly could," he says. "What I had were people around the table who were sympathetic to what I was saying but basically feeling tied down by political implications. They were just very nervous."

He pauses only briefly before adding, "I was told to repudiate some of it." At one of those meetings with Clinton staffers, Varmus was asked to endorse the more acceptable conclusions of the report—such as the funding of research on excess IVF embryos—but to draw the line at using government dollars to create embryos solely for research purposes. The thinking was that creating embryos just for scientists to study might be too disturbing for voters. Varmus refused. In his view, the worries about the creation of embryos solely for research purposes were driven more by political than scientific or moral concerns.

It was an important lesson for Varmus—and a harbinger of things to come for stem cell scientists—on the powerful role that politics would play in dictating how stem cell experiments were done. It's hard, in fact, not to think that Clinton's swift and decisive rebuttal of the panel's report played some role in fueling misconceptions about how and why embryo research was needed, particularly in the by-then conservative-majority House of Representatives in Congress.

Taking advantage of the favorable political climate, in 1996 members of the Right to Life movement approached Representative Jay Dickey of Arkansas and convinced him to write legislation that followed Clinton's lead and cemented into law a ban on using federal funds to support research on embryos. "The more I got into [the issue] the more I realized that what we might be doing is creating an industry of creating embryos maybe for profit," he says about embryo research. "That really bothered me, that people would be paying other people to have embryos created. My goal was to not have any termination [of embryos] no matter what the source is, or when in the process."

Dickey, then a junior member of the House, enlisted Senator Roger Wicker of Mississippi to cosponsor the bill, and in a clever political move, exploited a well-established tactic of lawmakers who wanted to insulate their bills from future manipulation. They attached the legislation as a rider to the yearly appropriations bill for the NIH. In effect, legislators who opposed the rider could not, and still cannot vote against it without voting against the budget for the nation's largest biomedical research agency. Passed into law by a predominantly conservative, antiabortion legislature, the Dickey-Wicker amendment ensures that no federal dollars are used to support IVF studies in which the embryo is harmed or destroyed, as well as, in a forward-looking act possibly triggered by the ill-fated HERP report, experiments on human embryonic stem cells. That possibility would occur sooner than any scientist or politician could predict, in a small room tucked away at a Midwestern hospital. In the meantime, the rider remains a part of the NIH appropriations bill more than a decade and a half later, renewed, along with the institute's budget, year after year.

4

From Mouse to Man

There is a room on the sixth floor of the University of Wisconsin hospital that sits empty now. To reach it, you must walk past the reception desk and waiting room for the Obstetrics and Gynecology clinic, past the Generations Fertility Care services, where couples come to take advantage of reproductive technologies such as in vitro fertilization, and past the active research lab attached to the clinic. You have to walk past all of this to the end of a long hallway of examining rooms to reach this room.

Its most renowned occupant has long since vacated, leaving only the barest traces of his presence behind. Other tenants have since come and gone, their only legacy some empty boxes that used to hold the necessities of biological research—pipettes, vials, and other supplies. Some drawers, previously filled with the business of a busy research lab, now sit half open, devoid of their contents. Someone is starting to use the narrow, galley-like space for storage.

There is an empty spot on the bench where a microscope for looking at tiny clumps of cells used to sit. The state-of-the-art ventilating hood is still there, and if someone were to turn it on, air would be sucked continuously up into an exhaust system to prevent contaminants from floating inadvertently down into the plastic plates of cell cultures that formerly buzzed in and out of circulation daily. But no one has turned it on in months.

The people who walk by this abandoned room see it as no different from any other lab in any other research department at any other university. Rumor has it that it will soon be gutted and renovated to become part of the expanding fertility clinic.

But if you ask the right people—and there are only a few who would even know—they will tell you, with just a little bit of awe and quite a bit of pride, about what happened in that space back in 1997 and 1998. It was there that a quiet, introspective man used to make a daily pilgrimage in the early morning hours to conduct a remarkable experiment. It was here, in this twenty-four-by-nine-foot room and under this now idle hood, that developmental biologist James Thomson nurtured a history-making set of cells. This was where Thomson managed to grow the very first line of human embryonic stem cells.

★ ★ ★

James Thomson is a ferociously private man, one who values the relative obscurity and routine of basic science over the often unforgiving spotlight of a high-profile career. Especially reticent about the media, he hates having his picture taken, grants very few interviews, and when he does, he tends to give sparse, to-the-point answers, taking care not to waste any words. Ask about his wife or family and you risk raising his ire over the unwelcome intrusion into his private life. He does not own a television, and is so reluctant about talking in front of a camera that he's done it only twice, despite the dozens of requests that have flooded into the university's public relations office from all over the world. (He resisted even personal pleas from the late Peter Jennings to be interviewed in prime time about his achievement.)

It's not that he has any problem with the press. It's just that he's not impressed with its track record in helping the public to understand stem cells. He begins our conversation in his small office on the University of Wisconsin–Madison campus by pointing out that despite all the

coverage of stem cells in newspapers and on news shows and documenta-
ries, most people still aren't sure where embryonic stem cells come from
and exactly what they can do.

"As a scientist, I want to change that," he tells me. "But my day job is
what I'm good at, so how much do I spend doing that, and how much do
I need to do press?" It's a question he clearly continues to wrestle with, as
his naturally insular tendencies clash with the very real responsibility and
obligation he feels to explain his work and inform the public about the
stem cell research that could one day change how we treat disease. He
remains selective in the interviews he gives, having come to the conclusion, he says, that "the press informs people but does not educate them."

It was a lesson he learned when the world suddenly appeared at his
door in November 1998, after he announced he had successfully generated the first batch of embryonic stem cells extracted from a human
embryo. As an academic scientist, he expected that with the news of
his achievement, people would ask a few questions and his job would
be done. From there, it would be the media's job to give the public its
crash course in developmental biology and hash out the complicated
and important ethical questions about how to use the remarkable biological gift he had given the world. But in retrospect, even Thomson
realizes how naive that expectation was. For those who know him, it's
the ultimate irony that such a singularly private man suddenly found
himself the poster person for the controversial field of embryonic stem
cell research.

Thomson didn't start out to give human developmental biology such
a jump start. As a graduate student at the University of Pennsylvania, he
trained in veterinary medicine and developmental biology, and given the
state-of-the-art in the field at the time, that meant focusing his attention
on the mouse. "It's sensible, because the mouse is relatively small and
easy to use," he tells me during our unusually lengthy talk.

Formerly at the University of Wisconsin's Primate Center, Thomson
is now based on the fourth floor of the genetics and biotech center, one
of the new additions to the still-growing campus's network of science

research buildings. The room is both small and spare in decor—a fitting reflection of its occupant, but a bit unexpected given his status, not just at the university but in the field of biology. For all the accolades and recognition he has received for his breakthrough achievement, the walls are noticeably bare of plaques and awards; this is clearly a man who does not flaunt his accomplishments. The most remarkable adornment in the room is a large fishbowl the diameter of a modest pizza tray, which serves as home to several dozen tiny guppies. "I've had that since I was in grad school," Thomson explains.

As he settles in for our discussion, he props his feet up on an extra chair that he keeps next to his own, seemingly for just that purpose. He picks up his story, and it's clear that he is driven, like many scientists, by a practicality that dictates his choice experiments. It didn't take Thomson long to realize that while the mouse was convenient, it wasn't an ideal model for understanding the most basic and important questions about human development, and especially about those critical first hours and days following fertilization. For one thing, key compounds that define important stages in the developing human embryo, such as the hormone human chorionic gonadotropin, which is what pregnancy tests measure to provide a positive or negative result, don't exist in the mouse. For another, mouse placentas are different from human placentas in the compounds they produce to cultivate the growing embryo.

So he turned his attention to primates, the closest evolutionary models for humans. But while monkey embryos were a better stand-in for studying development, they were nearly impossible to obtain. With one in vivo–produced embryo costing around two thousand dollars, it was nearly impossible to conduct the dozens of different experiments that Thomson had in mind.

"I was interested in the very earliest events in development, and it was because of that that I decided to make primate embryonic stem cells," he says of the decision that launched him into that world. "It was because I wanted to have a better, more accurate model of human development than was available with the mouse. And because the primate material was limited, I thought stem cells made a lot of sense, because you could grow as many as you want in quantity, and study them biochemically."

It was the early 1990s, and the only stem cells that anyone had successfully isolated from mammals were the ones that Gail Martin and Martin Evans had extracted from mice a decade earlier. And while most of the scientific community pounced on them as a stepping stone for generating genetically engineered mice lacking certain genes that they wanted to study, Thomson was among the handful of biologists who seized upon the cells as a window into something far grander. For them, the cells were the ideal resource for studying the earliest steps in development, to understand how a single cell—the egg—morphed and transformed itself into the multicelled, multifunctional network of tissues and organs of a fully formed animal.

"Yeah, I wouldn't say it killed developmental biology," he says of the rush to generate the knockout mice missing specific genes that helped scientists to learn what function certain genes had. (When researchers first identified the cancer-promoting p53 gene, for example, they knocked it out in mice to establish that it played a role in controlling cell division; indeed, without it, tumors in the animals grew with abandon.) "Knockout mice are hugely important. But a bunch of people doing really cool developmental biology shifted [their focus] and a lot of basic [developmental] stuff kind of got lost for a while," he says.

On the human side, researchers were beginning to appreciate a more limited version of the embryonic stem cells' pluripotent potential. Adult stem cells, as their name implies, are the mature version of their embryonic predecessors—if embryonic stem cells represent a blank page of biological possibility, the adult stem cells are more like already written chapters. They can be revised to a certain extent, but they can never be completely erased and returned to an entirely incipient state. So whereas embryonic stem cells have the ability to develop into any of the body's cells, adult stem cells are restricted to morphing into only a designated set of tissues—the various blood cells, for instance, for bone marrow stem cells, or for a heart stem cell, the various types of cardiac muscle.

And while they are useful for studying these limited tissues, adult stem cells were not as handy for developmental biologists like Thomson, who were interested in teasing out the first choreographed steps an embryo takes in growing from a single cell into a multicelled and complex individual. In addition, it wasn't clear that every cell type in a fully grown individual even had an adult stem cell it could claim as a progenitor. And even in cases in which such a mother cell existed, these adult progenitors were proving to be rare and devilishly difficult to isolate.

In order to study the black box of development from its very beginning, Thomson was convinced that embryonic stem cells would be the answer. He made a calculated bet that given how mouse researchers had plucked them from mouse embryos, it would be possible to do the same with primate embryonic stem cells. Or at least it should be.

Since Evans's and Martin's work, however, a string of scientists had tried to derive embryonic stem cells from other species—and failed. It wasn't clear whether the mouse embryonic stem cells were simply an anomaly that was particular to that species, or whether, as Thomson believed, it should be possible to extract the embryonic stem cells from any mammal.

Thomson is a careful scientist, and his confidence wasn't simply scientific bravado. As a graduate student, he had logged many hours in the lab of Peter Andrews, a biologist then at the Wistar Institute, in Philadelphia. In the 1980s, Andrews was working on the human version of the wild child of the stem cell family—the embryonal carcinoma (EC) cell that LeRoy Stevens had stumbled upon in mice at the Jackson Lab. Like the teratocarcinomas, the embryonal carcinoma cells would grow into anything and everything on a lab dish—they were unpredictable and unmanageable. But for eager young scientists like Thomson, the EC cell lines were the springboard for thinking about what would keep normal embryonic stem cells in the same state of perpetual youth. Andrews was focused on identifying the proteins that studded the surface of the EC cells, hoping to pick out the factors that kept the cells in their tumor state, replicating over and over again without changing into other types of tissue. It was the perfect proving ground for Thomson. "It just happened that I was in the same lab while that stuff was being developed, so I knew precisely how he cultured those," he tells me.

With that "stuff," Thomson was able to fashion the critical bridge between the carcinoma cells and embryonic stem cells. Borrowing heavily from the Martin and Evans approaches, he removed a six-day-old blastocyst from a fifteen-year-old rhesus monkey. Blastocysts are essentially hollow spheres of cells that contain two primary layers—the outer trophectoderm, which eventually becomes the placenta, and the inner cell mass (ICM), a cluster of cells clinging to the inside of the sphere that make up the embryonic stem cells. It's the inner cell mass that eventually morphs into the multicelled growing fetus. Those were the cells that Thomson was after.

To remove them, he relied on a clever technique first developed in the 1970s on mouse embryos. Called immunosurgery, it exploits the Velcro-like ability of antibodies to stick to specific proteins. In Thomson's case, he used antibodies that adhered to the outer trophectoderm cells, and then killed these attached cells with a cell-lysing agent. Once he washed away the debris, he was left with the inner cell mass. He then carefully transferred these surviving cells to a specially prepared cocoon of growth factors, nutrients, and mouse skin cells (which, as Gail Martin had discovered, release important factors that keep the stem cells healthy) on plastic plates. After nearly six months, he finally had colonies of cells that looked similar to the ECC lines he had grown before. The primate cells were sprouting the same proteins that the immortal EC cells did, and they continued to proliferate without forging ahead to form more mature cells. They remained, as Thomson wanted them to, in their embryonic, or undifferentiated state.

As a developmental biologist, Thomson was excited by his success. Or as excited as the reserved scientist gets. While he rarely expresses his enthusiasm publicly, he tends to speak volumes in his papers. In the five-page publication describing his work in *Proceedings of the National Academy of Sciences* in August 1995, Thomson titled his experiment simply "Isolation of a Primate Embryonic Stem Cell Line." And while he was all business in describing his meticulous methods in the experiment, in the final two paragraphs of the study, he allowed himself to get a bit more expansive as he laid out the beginnings of the stem cell field to come. "Human ES cells would offer exciting new possibilities for transplantation medicine," he wrote. "Because ES cells have the developmental potential to give rise to all adult cell types, any disease resulting from the failure of specific cell types would be potentially treatable through the transplantation of differentiated cells derived from ES cells. . . . Because of the range of disease potentially treatable by this approach, elucidating the basic mechanisms controlling the differentiation of primate ES cells has dramatic clinical significance."

It was the closest that Thomson would come to blaring the trumpets and declaring the dawn of a new age in developmental biology. In fact, in his own way, Thomson was predicting a revolution in medicine as well, proposing an entirely new way of thinking about diseases and finding methods to treat or even cure them. Instead of merely addressing symptoms of chronic conditions such as diabetes or Parkinson's, here was a way, he said, of potentially replacing troublesome cells with entirely new, healthy ones.

Oddly enough, however, Thomson's declaration seemed to fall on deaf ears—at least among the media. An Associated Press reporter who happened to be attending a media workshop on the Wisconsin campus at the time learned of the experiment from the proud chair of the Primate Center at a cocktail reception. He rushed off to file the story, but even his article ran in only a handful of papers. And the story focused on the utility of stem cells more in enhancing gene therapy—which was then the hot science—than as a truly game-changing tool for treating disease.

"Nobody cared," Thomson says, "except for the little community of scientists who were studying human ECCs. They got it, and said, 'Oh wow, look.'" Then he adds, revealing his wry sense of humor with a smile, "So that would be six people in the world."

Perhaps it was the fact that the work had been done in monkeys, who, despite being our closest evolutionary cousins, are still several branches away from *Homo sapiens* on the lineage tree. Or perhaps it was simply too novel an idea—the concept that a source of primitive, still immature cells could serve as a theoretically unending depot of new cells and tissues to repair or replace damaged ones.

But Thomson and that handful of developmental biologists knew. And in his primate lab in Madison, Thomson was convinced that someone else working on the human side of things would take the logical next step and extract the same stem cells from a human embryo.

"It surprised me that more people didn't jump on it, yeah," he tells me. "Several people tried, and didn't succeed for whatever reason. But there weren't that many people interested in the rhesus paper, so not a lot of people noticed it. We thought somebody else would do it, but as the months ticked by, and nobody else had done the human work, we decided to pursue that," he says of the why-not decision that would lead to his pioneering human embryonic stem cell lines.

As with the primate work, in which Thomson benefited from his experience with growing ECC lines, he would now have the advantage of having worked with the primate cells, and draw upon his familiarity and experience with both types of lines. The issue with his new experiment, however, would be resources.

The key to his success with the primates had been his easy access to the rhesus embryos. As it happens, the University of Wisconsin's Primate Center is one of eight national primate research facilities in the United States receiving federal funding. Human embryos, however, would involve additional layers of oversight and review. Before he could even touch his first human embryo, Thomson realized that the experiments he wanted to do involved a thicket of ethical and moral issues that would have to be addressed outside of his lab.

It was 1995, and in Washington, concern over just the type of work that Thomson was proposing had prompted Congress to pass the Dickey-Wicker amendment, making human embryos all but toxic for future government-funded researchers. No one was sure yet how rigorously government officials would be in auditing studies involving human embryos, but few were willing to find out. Since the amendment prohibited the use of government dollars toward any studies that involved even the harming of human embryos, Thomson would have to find financing elsewhere. Because in order to extract stem cells, the embryos would have to be destroyed.

"I didn't have money, basically," he says. "I could not work at the Primate Center because that was a federally funded facility. So I tried to figure out how to get funding. There was a pot of money at the university that I tried to get, and I think they were afraid, so they said no."

That pot was earmarked specifically for non-federally funded projects like the one Thomson was proposing, but the experiment skirted close enough to the limits of federal regulations that the university was understandably gun-shy about investing in the studies, fearing that the millions it was receiving from the National Institutes of Health would be in jeopardy. The university is regularly among the top twenty institutes in NIH grant awards each year.

That was when a man named Michael West walked into Thomson's office and made an offer that the scientist simply could not refuse. West, the brash and forward-thinking founder of Geron, the biotech company that at the time focused on antiaging technology, was quick to realize that stem cells represented a potentially new way to regenerate human tissues and, by extension, human life. He had been sent to Madison at the recommendation, just a few days earlier, of Dr. Roger Pedersen, a biologist at University of California, San Francisco.

Pedersen had become interested in human stem cells by way of his work in improving IVF techniques, and was captivated by news of a Singaporean scientist's attempt to culture human embryonic stem cells from an IVF embryo. Ariff Bongso had come close to becoming the first person to accomplish this feat, but he couldn't get the inner cell mass to survive on a culture dish. Pedersen thought he could.

But when West first approached him with an offer to finance his attempts, Pedersen says he "essentially said no way. I thought that if we were going to achieve this, it was something that should stay in the public domain, and it shouldn't be tied up in patenting and by a company. West said, 'Well, if you ever change your mind, let me know.'"

Pedersen didn't think he would, but that was in 1992, before Dickey-Wicker, and he could afford to believe, if naively, that the federal government would sponsor stem cell studies. Once that legislation was in place, however, he was obliged to do an about-face. He called West and accepted his offer. West, for his part, shrewdly realized that if the principled Pedersen was desperate for funding, others would be too, and as a biotech executive, he saw an irresistible opportunity to mop up this orphaned innovation and wring it into potential market dominance for Geron. Based in nearby Menlo Park, West came back to the university's Parnassus campus and asked Pedersen a critical question: Who else was working on generating human stem cells?

Aware by that time of Thomson's paper on creating the primate stem cell line, Pedersen told West about Thomson and another biologist on the East Coast, at Johns Hopkins University in Baltimore, by the name of John Gearhart. "I have always been very impressed by this decisiveness," Pedersen says with a laugh, "but Mike was on a plane the next day to Wisconsin. And he eventually signed up Jamie into a support agreement which had intellectual property implications as well."

West's timing couldn't have been better. It was a week after Thomson had been turned down by the university. And while he, like Pedersen, was reluctant to have such seminal work funded by a private entity, Thomson was also backed into a financial corner. "He basically asked me do I want to do it or not, and since the university said no, and the government said no, this seemed like the only option at the time," Thomson recalls, before rolling his eyes. "It's unfortunate that we weren't given a broader choice than that."

Yet despite now having the money to do the experiments, it wasn't an ideal situation. Because his own lab was supported by government dollars, the first thing Thomson had to do was find a space, preferably on campus, that wasn't financed with federal funds—no small feat. Thomson finally settled on an unused space on the sixth floor of the university hospital. Now he had the money to support his project, and a place to perform his experiments. All he needed was permission to do them.

★ ★ ★

Before a scientist can conduct any experiment on a university's grounds, he must clear his study with a review committee known as the institutional review board, or IRB. To ensure that volunteers are not coerced, exploited, or intentionally harmed by a proposed experiment, IRB approval is necessary for any study involving human subjects.

But the project that Thomson was proposing wasn't just any study involving human subjects. It involved embryos. Would they qualify as human subjects? Who could give consent to study them, and how would he obtain that permission?

Aware of the political minefield he was about to walk into, one of the first calls Thomson made was to Alta Charo. Having read the Human Embryo Research Panel's report from the previous year, he knew Charo would have a firm grasp on the legal, ethical, and political storms he would need to navigate. And it didn't hurt that Charo happened to be on the law faculty, just a few buildings away from the Primate Center. Eager to proceed with the human studies, but wanting to be fully aware and respectful of the regulatory requirements to do so, he asked Charo to come to his labs for an introduction and discussion. "I wanted to know what I was getting myself into," explains Thomson.

"I went over, and first he gave me a seminar on embryonic stem cell research, which was terrific," recalls Charo of that initial meeting. "We talked for quite a long time, and more than once after that, about the legal issues and the various ethical analyses out there, and the political reality out there about doing this kind of work."

On the broadest level, Thomson was concerned about the ethical frictions his experiments would raise. After Dolly, it was clear that the American public had been caught off guard and was still struggling to understand the full impact of what it meant to now have the power to turn any mature cell into an entirely new animal. While Thomson had no intention of cloning a human being, he was wading into the same moral soup that Dolly's existence had stirred up. Although he was proposing to use existing embryos—those that couples at IVF clinics would donate to his research after deciding they no longer needed them for in vitro fertilization—they were still embryos. Theoretically, if thawed and implanted into a woman's uterus, they could develop into a baby.

The lawyer and the scientist soon found themselves ensnared in a thicket of complex regulatory questions. Were human embryos considered human subjects? Was there something special about the embryo, before it was implanted in a womb, that exempted it from human subject regulations? If the work were privately funded, would the government's regulations dictating how human subjects were to be treated still apply to the embryos?

"For somebody who was a scientist, it was completely unfamiliar and fairly complex," says Charo. But Thomson was willing to learn, and committed to doing it right. His intent was not to make a political statement with his experiment; he was genuinely interested in studying human embryonic stem cells for their ability to answer some pressing questions about early development. And the last thing he wanted was to jeopardize his ability to do the science by stumbling on the regulatory requirements for setting up the study. "From the very beginning, his goal was not to make a political point," says Charo. "He was truly interested in the basic science of human embryology. He really wanted to understand what makes the embryo grow normally and not normally. He had reached the limit of what he could do on primates in order to understand the human embryonic development process."

At some point in their discussions, Charo advised Thomson to think about whether he was personally comfortable with the experiment he was proposing: Putting aside all the regulatory challenges they were uncovering, was he comfortable with studying human embryos in the way he was outlining? Years later, Thomson remarked to a reporter who asked him about the ethical controversies surrounding his work, "If human embryonic stem cell research does not make you at least a little bit uncomfortable, you have not thought about it enough."

"Yeah, I know the pro-lifers liked that one," he says with a laugh. But his intention, he adds, was to highlight that the ethical debate couldn't be accurately conveyed in the black-and-white arguments that were appearing in the press. Not all scientists were amoral technocrats with little respect for human life, and by the same token, not all conservatives who were against abortion were necessarily against the proper and responsible study of the human embryo.

"The people for whom I have the most respect on this issue are those who you would think would be knee-jerk opposed, but actually took the time to think about it," he explains. "Is it ever acceptable to kill a human embryo? In a sense you can view it fairly simplistically. In current medical practice, those embryos are discarded [if a couple undergoing IVF no longer needs them]. It's a separate question whether or not that is a good idea. We don't have an answer, but as long as you are not complicit in that decision, isn't it a better ethical outcome to use those embryos for something that helps people?"

After two years of discussions, the University of Wisconsin's IRB finally agreed. It approved Thomson's study, but that didn't mean university officials weren't still nervous about the unprecedented experiment. And that's how Thomson found himself in that tiny room behind the infertility ward at the hospital.

Although Geron would be funding his studies completely, the academic administrators weren't sure how government officials would view intermingling that private funding with federal grant money that was already covering more general expenses such as utilities (electricity and water) as well as computers and other sundry supplies that are the mainstay of any research facility. Would such coexistence run afoul of Dickey-Wicker?

To avoid any doubt, and avoid any hand-wringing by the university or any potential missteps, Thomson decided it was better to be safe than sorry. Picking up discarded equipment—"junk, really"—that had not been part of any government grants, he stocked the tiny lab with the barest necessities for deriving cells. He also decided to work alone, because, he says, "I derived all the primate cell lines myself, and I was the one who knew how to do it. I did the original work over at the hospital in a little dedicated room sequestered from everybody."

★ ★ ★

Thomson's plan was to rely primarily on excess embryos from the adjoining IVF clinic. But while the staff there were willing to approach patients to donate embryos, the Madison facility wasn't exactly doing a high-volume business. And for all Thomson knew, he might need hundreds of embryos before he would be able to successfully nurture a stem cell line from one of them to grow.

In another case of having the right person walk in at the right time, it happened that Thomson had been hearing from a gynecologist at Rambam Medical Center, in Haifa, Israel. Ever since reading Thomson's paper on isolating rhesus embryonic stem cells, Joseph Itskovitz-Eldor couldn't shake the idea of translating that technique to human cells. As a fertility specialist, he could appreciate the potential of using embryonic stem cells to better understand human development. And he had access to IVF embryos. Persistent, with an unmistakable gravelly voice, Itskovitz-Eldor had spent time at the Jones Institute for Reproductive Medicine, in Norfolk, Virginia, where he learned the technique of forcibly microinjecting sperm, rather than allowing them to fertilize eggs on their own in a dish, to improve fertilization rates of IVF embryos. He brought the technique back to Israel and became the first to use it at the Rambam clinic. With embryonic stem cells, he thought, he might learn more about the earliest steps in human reproduction; he pursued Thomson relentlessly for a collaboration. But Thomson, more of a loner than a joiner, was unsure about whether a partnership on such a controversial project made sense.

Working through a mutual acquaintance who was studying the cloning of cows, however, Itskovitz-Eldor finally managed to invite Thomson to visit him in Israel. "He came for two or three days to the Holy Land, we visited the Jordan River, and I asked him again to collaborate," he says. After a few more letters back and forth, Thomson finally agreed. Itskovitz-Eldor sent a series of frozen embryos, along with one of his students, who stayed with Thomson and his wife while learning cell culturing, to Madison.

Thomson now had two sources of embryos. All he had to do was figure out how to tease apart their embryonic stem cells and keep them alive in a culture dish.

That didn't take long. "I always thought that if we had access to decent-quality embryos, then essentially we could do it right away," says Thomson. In the end, the preparation took longer than the experiment itself. Within a month of getting his hands on the first embryo, from the Wisconsin hospital's IVF clinic, Thomson had his first colonies of stem cells sprouting on his plastic plates. In fact, Wisconsin's cell lines are numbered H1 to H14; H is for "human," and the numbers designate the order of the embryos that were used in the study. The very first embryo Thomson used, H1, yielded a valid stem cell line.

His experience isolating the monkey stem cells ended up guiding his recipe for culturing the human cells. After carefully thawing the first embryo from its deep freeze in liquid nitrogen, he cultivated it on a dish in the lab using a similar mix of nutrients and growth factors that IVF clinics use to keep the embryos alive for four or five days, to blastocyst stage. He did the same for thirteen other embryos, some of which had been brought from Israel by Itskovitz-Eldor's technician.

Using the same immunosurgery technique he had applied to the primate blastocysts, Thomson then washed out the inner cell mass, or ICM, from the human embryos and plated them on his mix of mouse feeder cells, growth factors, and media. Between nine and fifteen days later, he picked out the first clump of outgrowths that would become embryonic stem cells.

As Thomson learned from his previous work on monkeys, if the cells were happy, they would start to grow quickly, consuming their nutrient broth and rapidly out-spacing their confining plastic homes. That's why he had to inspect them once a day, every day, to make decisions about when to feed them and when to perform a pick-and-split.

Picking and splitting, as anyone who later learned to grow these embryonic stem cells quickly discovered, is a tedious and inexact endeavor, more art than science. Every few days, as clumps of cells start to pile on top of each other, the outgrowths need to be manually separated with a pipette and transferred to another dish.

Rogue cells that manage to break free from their embryonic stasis and sneak out to develop into a more specialized type of cell also need to be picked out and removed. And it isn't uncommon to see quite a few such outliers—when undifferentiated embryonic stem cells overpopulate a dish and deplete their food source, the competitive pressure for nutrients pushes them to start differentiating, since that's their path of least resistance. Maintaining an undifferentiated state is a highly unnatural and high-energy task—a type of suspended animation. Given any opportunity, the cells attempt to break free from their stasis and give themselves an identity—in most cases, embryonic stem cells, it turns out, want to become neurons. But leaving the plates too sparse isn't the answer either, since the cells appear to be social animals—they need the proximity of others in order to feel comfortable enough to continue to divide. So Thomson's job was to keep the cells at just the right density to encourage them to grow but not to differentiate.

That meant feeding the cells a broth of nutrients, cow serum, and amino acids that he had mixed up to keep them happy. Normally a bright orange red, this medium would turn a dull yellow when cells had exhausted its nutritive powers. (Stem cell labs, like Thomson's, are known for their refrigerators, which house nothing but bottles of cheerily-colored liquids that resemble Kool-Aid.) To change the media, Thomson had to suck out the old fluid by mouth using a specially designed pipette (fancier labs are now equipped with an aspirator that vacuums out the liquid). The entire process has to occur under the hood, which pulls air up and away from the work surface to prevent any particles or pathogens from settling on the dishes—any unwanted foreign resident could contaminate and destroy the entire plate housing millions of stem cells.

For months between 1997 and 1998, Thomson made the trip up to the sixth floor to nurture his cells. Day after day, he would inspect them; every few days, he would carefully change the media, and almost every week he would pick and split the most aggressively growing colonies. The ICM from five of the original fourteen embryos continued to grow, and after several months, he became cautiously optimistic that what he was caring for and feeding were indeed human embryonic stem cells.

Generating the stem cells, as Thomson tells it, was the easy part. Verifying that the cells indeed remained undifferentiated, and contained normal chromosomes, was considerably harder. Knowing how much scrutiny both he and the cells would face, Thomson wanted to be sure that what he had generated were indeed embryonic stem cells, capable of differentiating into all of the body's cell types. So in addition to the criteria that researchers used to define undifferentiated embryonal carcinoma cells (at the time, they were the only other pluripotent human cells known to exist), Thomson imposed his own requirement. He maintained his cells in their suspended embryonic state for nearly a year.

On the day he stained his newly formed cells with Vector Red, a deep purple fixative that sticks to undifferentiated cells and turns them blue, he pulled a nurse from the adjacent IVF clinic who happened to be walking by into his tiny lab and asked her to look into his microscope. "He just asked me what I saw," Lynn Boehnlein recalls. "I looked in and said I saw something that was blue." That was apparently the right answer, and Thomson immediately asked her to sign his lab book attesting to that fact. But what stands out in her mind was Thomson's mood after she reported what she had seen. "He was giddy, really," she says, "which was so unlike him." What Boehnlein couldn't have known was that the cells she saw were saturated with alkaline phosphatase, a hallmark of embryonic stem cells. After Thomson, she was the first person to see stem cells extracted from a human embryo and nurtured on a lab dish.

★ ★ ★

That blue stain, however, was just one of a series of tests that Thomson then spent the next six months conducting on his precious cells. "Having stem cells is one thing, but to have viable stem cells you have to set them through their paces—you have to freeze them and show that you can thaw them out again. Because if they die, then nobody is going to believe you."

Thomson's checklist mirrored the one that he had relied upon to establish the identity of his monkey embryonic stem cells: The cells had to be frozen and then thawed and coaxed to grow again; they had to form a teratoma that contained cells from all three basic germ layers (which go on to form the more than two hundred cell types in the human body)—endoderm, mesoderm and ectoderm; and they had to test positive for proteins that were unique to undifferentiated embryonic cells.

Finally, he simply wanted to prove that the cells would remain in their pristine state without marching off and developing into nerve cells or skin cells or any other mature cell types. That, it turns out, wasn't a trivial thing, since the differentiated state is the default, and cells want to turn themselves into something more developed, rather than stay in a sort of embryonic suspended animation. "Embryonic stem cells themselves are a tissue culture artifact," Thomson points out. "The same cells in the embryo itself are really not embryonic stem cells but a precursor cell that represents only a very transient state."

Given the uneasy team of rivals that Geron had assembled—Thomson, Pedersen, and Gearhart—all of whom were racing to culture the first human embryonic stem cell line, Thomson wanted to be extra sure he was right. At one point, during a visit to San Francisco, he was characteristically blunt with Pedersen. "He said, 'If I had them myself, I wouldn't tell you,'" Pedersen says with a laugh. What Thomson didn't reveal to him then, but admitted later, was that he had actually seen embryonic stem cells in Pedersen's dishes but didn't tell him. Years later, Pedersen doesn't seem upset over being upstaged, ceding that Thomson knew something he didn't about keeping the human cells alive—it turns out that unlike the primate cells, which were happier in smaller clumps of five to ten cells, the human cells were far more needy, thriving better in larger bundles. "They had the advantage of knowing the effect of clump size on the viability of the rhesus cells, and having the intuition to move to larger clumps," Pedersen acknowledges.

This time, Thomson got noticed. Having watched the press descend upon the small town of Roslin following Ian Wilmut's announcement of Dolly's birth, Thomson had some idea of how his quiet existence in Madison would change. "I knew precisely what was about to happen," he says, laughing now about the media circus that ensued. "The feeding frenzy had already kind of begun with [Wilmut]. I knew they would just switch to me. And they did."

His three-page report, published in *Science* in the first week of November 1998, included those images of the blue-stained embryonic stem cells that Boehnlein had seen. Thomson's results were dissected and analyzed across the front pages of nearly every major newspaper. And this time, his prediction of the potential for embryonic stem cells to transform medicine didn't go so unnoticed. "Many diseases," he wrote, echoing his summary from his 1995 paper, "such as Parkinson's disease and juvenile-onset diabetes mellitus, result from the death or dysfunction of just one or a few cell types. The replacement of those cells could offer lifelong treatment. . . . Progress in basic developmental biology is now extremely rapid; human ES cells will link this progress even more closely to the prevention and treatment of human disease."

Within a month, Thomson found himself blinking in front of a microphone in a cavernous Senate hearing room on Capitol Hill, flanked by Harold Varmus, from NIH, and John Gearhart, who, thanks to Geron's support, had also isolated human embryonic stem cells, except from fetal tissue rather than embryos. The scientists were simultaneously introducing the nation's representatives to stem cells and defending the very existence of the yet-to-emerge science behind it. Conservative and pro-life groups had already seized upon the source of the cells—embryos—as morally unacceptable.

"I had no idea what it would be like," Thomson says of his first appearance on Capitol Hill, at a hearing chaired by Pennsylvania senator Arlen Specter. "I was scared to death because I did my graduate work in Philadelphia and I knew who [Specter] was, and I remembered the Anita Hill stuff and I thought, 'Great, I'm going to be like Anita Hill.' But it turned out he was very respectful, and asked very good questions and understood it very well."

Even then, however, Thomson was keenly aware that the purpose of the hearing was not entirely to celebrate the work and figure out ways to promote it, but to learn more about the ethical as well as scientific challenges that it posed. "I really had no idea what the reception would be— whether it was 'This is great for medicine,' or whether it was that I was doing evil work. I did feel going in there that I was defending whether we could do this or not."

Although Thomson had funded his experiment with private investments, he was making a plea to allow government money to flow into the field in order to help it flourish. Embryonic stem cells, he told the senators, could generate large populations of human cells, including muscle, pancreatic, and neuronal cells to treat disease. But with the Dickey-Wicker ban on NIH funding for human embryo research, he continued, stem cell studies would also suffer. He then endorsed the conclusion reached by the Human Embryo Research Panel, on which Charo had served in 1994, and supported its recommendation that "It is in the public interest that the availability of federal funding and regulation should provide consistent ethical and scientific review for this area of research."

Even now, with the benefit of hindsight, Thomson believes that the government's failure to invest in embryonic stem cell research, as the 1994 panel had advised, has slowed progress in the field. "It's clear that the political process inhibited the rate of progress in the field," he says. "That's unambiguous. From 1998 to 2001, there was no federal funding for this. In the United States, investigators were effectively excluded from participating because nobody had a lab to do this in."

Even if they did manage to raise private funds and set up a satellite facility or some type of temporary structure to house the human embryonic cell labs, investigators found themselves in a very uncomfortable position. Without federal oversight of the field, they were left to fend for themselves when it came to setting up their experiments, getting them reviewed and approved by their individual institutions, and, perhaps most important, gaining access to the cells themselves. By virtue of the fact that Thomson's breakthrough was financed entirely with private funds, the intellectual property and whatever licensing rights that went along with that ownership belonged entirely to him. Which meant that anyone who wanted access to the cells would have to go through Wisconsin.

* * *

Because the government hadn't decided where it stood with respect to human embryonic stem cell research, from Thomson's perspective the story quickly changed from one about science to one about politics. And patents. It was never his intention to become a stem cell scientist; for him, these cells were a detour along his initial journey to explore the early stages of human development. "When I first derived those original five cell lines, I figured, hey, there will be a whole bunch soon, and these [five] are not terribly important," he says. Exhausted by the bureaucratic and administrative mountains he had to climb in order to conduct his experiments, he wasn't planning on going through the process again. "I didn't want to make more, so I thought nobody would bother me," he says.

A scientist, not a businessman, Thomson decided to assign all patent rights to the university so that he could go back to the lab and leave it up to others to work out the legal and financial ramifications of his work. And as it happened, the University of Wisconsin has one of the nation's oldest, most lucrative, and, in the opinion of some researchers, most aggressive technology transfer offices in the country for just such discoveries.

The nonprofit Wisconsin Alumni Research Foundation (WARF) had been established in 1925 by three university professors who saw the value in patenting scientific discoveries in order to protect them from abuse and misuse. As the story goes, in the late 1890s Stephen Babcock, professor of biochemistry, discovered a way to test cow's milk for its butterfat content. Believing that the technique should be available for any farmer and consumer to use, he made the fateful decision not to patent his invention, and before long farmers and cow dealers were abusing the technology to manipulate the butterfat content of milk sold on the market.

So thirty years later, when Harry Steenbock found a way to capture vitamin D's nutritive benefits and embed them in foods such as milk, he wanted to avoid the same misapplication of his invention. He and two colleagues created WARF, which filed for the patent on the technique using three hundred dollars of Steenbock's own money; it remains the university's most profitable technology to date, pulling in several billion each year and still accounting for 60 to 70 percent of the foundation's revenue.

When Thomson disclosed his work on the primate stem cells to WARF in 1994, the foundation immediately filed a patent on the process and the resulting cells. When Geron's initial one million-dollar investment in Thomson's work ran out, WARF contributed additional funds to fill the gap. That patent would be followed by two more, in 1998 and 2001, to establish its claim over the human versions of those cells as well.

In the coming years, WARF would be seen as the eight-hundred-pound gorilla in the stem cell world—it was a metaphor one scientist used in frustration to refer to the foundation, and it stuck. WARF made decisions that to many in the field smacked of monopoly and even greed. Initially, it charged researchers five thousand dollars for two vials containing a million or so embryonic stem cells each, an exorbitant sum at the time for cells of any kind. And because Geron had contributed to Thomson's studies, WARF was also obligated to honor a licensing agreement that the university and Thomson had agreed to, in which the biotech company enjoyed exclusive rights to developing the stem cells into six cell types: blood, bone, pancreas, liver, muscle, and nerve. This would then enable Geron to turn these cells into possible treatments for the most common chronic diseases for which there were still no cures, including diabetes, Alzheimer's, Parkinson's, and ALS.

The agreement proved to be a sore point for stem cell scientists and eventually for WARF, who ended up suing Geron over how far its licensing power extended. It doesn't take a medical expert to realize that Geron had skimmed the cream of the disease crop as far as new treatment opportunities were concerned. What, complained many scientists, was left for them to develop? With Dickey-Wicker effectively chilling government grants in the field, which private investors would sink their precious research dollars into an area in which their potential payback for developing and eventually commercializing discovery was so small?

It was just the acrimonious and energy-sapping situation that Pedersen had predicted when he initially turned down Michael West's financial support. With private concerns calling the shots, he and others believed, science would not get done the way it should. Because Geron and WARF had provided all of the financial support for the derivation of the cells, and because WARF now owned the patent for them, the foundation had every legal right to dictate what could be done with them. They could treat them as proprietary material, and keep them sequestered in their own labs to exploit for their own financial gain, or they could make them available to other researchers—under certain conditions. Either way, WARF was in control, and any investigator who wanted to study the cells had to agree to the research foundation's terms.

And those terms, at least to start out with, weren't exactly attractive for anyone outside of Madison.

★ ★ ★

"Our first mission is to promote, encourage, and aid scientific investigation and research at the University of Wisconsin–Madison," Carl Gulbrandsen, managing director of WARF, says by way of introducing me to the foundation. "We serve the university." It's apt, then, that WARF's offices occupy the top floors of the tallest structure on the university's Madison campus, in a hard-to-miss dark-glass fifteen-story edifice known as WARF tower, which stands sentry over Lake Mendota to the north and the school's campus to the east.

It was in the fifteenth-floor conference room of the tower, one floor above Gulbrandsen's office, that he and the foundation's board debated and eventually decided to patent Thomson's human embryonic stem cells in 1998. And it was that decision, claim many scientists, that set the stage for a contentious battle over access to the cells for research, and even, in the view of some, hampered the progress of stem cell science.

A trim, direct man with a neatly combed head of white hair, Gulbrandsen didn't make himself popular among stem cell scientists with his rigid adherence to WARF's Wisconsin-centric mission. He's not a man who needs to be liked, and even seems to enjoy piquing people with reminders of WARF's dominant position when it comes to embryonic stem cells. It doesn't take long for him to recount his first encounter, on a scientific panel, with Jeanne Loring, a scientist at the Scripps Research Institute, who informed him she studied human embryonic stem cells and had created a company based on her work. Gulbrandsen responded that she must have a license from WARF. "She didn't like that," he says, laughing.

But he admits that when Thomson first informed WARF of his break-through, "it wasn't a given that we should patent this. I remember in the discussion there was some consternation because these were human cells."

Patenting human tissue, because it already exists in nature, was a hotly debated topic, ever since scientists and institutes had begun claim-ing their stake to bits of the human genome. By 1997, when WARF began its discussions about human stem cells, thousands of genes had already been patented with the U.S. Patent and Trademark Office. Embryonic stem cells, some members of the licensing and patent staff argued, could not be patented because they were human material. Others, including Gulbrandsen, maintained that they essentially were not natural since they were grown in a dish, and, he says, "my feeling has always been that patent law is to a great extent agnostic, and it's up to legislators and others to determine whether it is a problem. If it's new, and it's novel and it's not obvious, then we should file an application on it."

But if WARF had decided not to patent the cells, I ask, wouldn't it have allowed more researchers to gain access to the cells and ultimately set up a more fertile environment in which the field could grow free of licensing fees and royalty concerns? After all, it was already being pum-meled by ethical and political forces; could WARF have alleviated some of that pressure by eliminating the legal and patent hurdles?

Gulbrandsen smiles. It's a question he has been asked many, many times before, and he has a ready answer. "Under our agreement with Geron, if we hadn't filed a patent on it, then Geron would have filed a patent on it. Who do you want to own a patent? A commercial company— then the researchers would really be in a good place," he says with a laugh, "or a foundation? I chose to think that if intellectual property is going to be owned, it's better that it be owned by a foundation that at least had a public-good mission at the end of it."

But even Gulbrandsen admits that the public-good mission was not entirely altruistic. Yes, WARF did indeed intend to distribute the cells in order to promote stem cell research, but in a way that was advantageous to the university, the protection of whose interests, as Gulbrandsen reminds me several times, is the foundation's primary mission.

The most pressing of these interests was WARF's obligation to Geron. In its pact with the biotech company, WARF had promised licensing rights to any commercial products or treatments derived from six cell types. In order to protect and honor that commitment, the foundation opted to patent the cells and the method that Thomson had used in creating them. The patents, as WARF sees it, were key to ensuring that companies like Geron would continue to be interested in stem cells, and would continue to back that interest with investment dollars. Without patent protection, Gulbrandsen argues, stem cells would not make good business. "If you don't have strong patents, it's hard to attract investment dollars," he says.

But what makes good business sense did not translate to equally sound science sense. For researchers, the patent proved to be an administrative headache. It meant that anyone wishing to get their hands on the cells, or make the cells on their own, would have to obtain a license from WARF in order to do so—even if they had no intention of commercializing their work.

And the terms of that license made more than a few scientists see red.

The materials transfer agreement (MTA) is a staple of academic science. Two institutions create an MTA whenever researchers from one university agree to provide cells, cultures, reagents, or other materials that they have developed to their colleagues at other universities. MTAs cover the use of these materials for research purposes, and normally the only costs involved are nominal fees to fund distribution and shipping. The idea behind them is to promote sharing of resources in order to foster innovation and expand on the original discoverer's findings. If the receiving researcher were to develop a marketable product, then the interested parties would negotiate a licensing agreement. But until that time, for purely research purposes scientists for the most part have free rein to conduct their own experiments without worrying about much oversight or restriction from the originating lab.

WARF's MTA for Thomson's stem cells, however, contained a few extra conditions that extended above and beyond the usual agreements.

For one, WARF strictly laid out what type of research the cells could and could not be used for: Under the foundation's criteria, they could not be combined with animal eggs to generate human-animal chimeras, nor could they be introduced into another human embryo or transplanted into a woman's uterus. And the most galling conditions: If the researcher ended up developing a new treatment or therapy that WARF wanted to commercialize, then the foundation was entitled to an unspecified portion of revenues or royalties from that. In addition, the foundation requested that scientists send in a yearly letter describing their studies as a way of ensuring that these conditions were being met. The agreement also gave WARF the right to terminate, at any time, any ongoing studies with the cells and ask for the cells to be returned.

Back in Boston, Doug Melton, who was eager to work with the cells, was dumbfounded by the terms of the MTA. He saw them as an insidious way of effectively making him an employee of WARF, by forc- ing him to provide the foundation with a yearly update of his work. Particularly irksome was WARF's right to terminate any experiment at any time it deemed the requirements were not being met. Melton called Gulbrandsen himself and directly questioned WARF's understanding of how science is done. "I said it's inconceivable that I could ask a postdoc or graduate student to work with this material, knowing that at any moment in the future, you would ask us to destroy this information and all the cell derivatives, and pull it back."

Gulbrandsen recalls the exchange, and acknowledges that he probably didn't address Melton's frustration. "I said, 'There's nothing I can do about this. These are the provisions we inherited, and we have to abide by them. If you don't like them, then don't use these cells.' That didn't please him," Gulbrandsen recalls with another smile.

Despite how it appeared, the conditions, he says, were not WARF's attempt to control or dictate the types of experiments that researchers could perform, but rather reflected the foundation's attempt to honor the informed consent that the couples donating the embryos had signed. In that document, the donors were assured specific things—that their embryos would not be used for certain types of experiments, such as those in which stem cells from their embryos would be transplanted into an animal egg to create a chimera. In addition, Gulbrandsen reminds me once again, there were the university's interests to consider. "We live in a very pro-life environment here in the state of Wisconsin," he says. "We needed some mechanism to say in fact we were taking these promises to donors, these prohibitions against certain types of research, very seriously." Requiring a yearly testament from scientists who received the cells, he says, seemed like an innocuous enough way to document that compliance. And the right to terminate any experiments that the foundation felt had violated these terms was also a way to reassure the state's legislature that it would not tolerate misuse of the cells.

The net effect, whether intentional or not, was to establish WARF as that eight-hundred-pound gorilla in the still tiny stem cell jungle, and to enhance the foundation's growing reputation as a profit-hungry, monopoly-seeking behemoth. It didn't help matters that publicly, at least, WARF officials made no apologies for their policies, and continued to aggressively enforce them, rather than adopting a more collaborative position (something they eventually did, but perhaps too late to rebuild their fractured relationship with the scientific community).

It was a cockiness that emerged from confidence. At the time, Thomson's cells were the only ones around, and as the political fight over whether the government should or should not endorse stem cell research began to take shape, it seemed very likely that Wisconsin would remain the only source for quite some time. Like it or not, the gorilla had settled in, and as far as anyone could see, wouldn't be budging anytime soon.

★ ★ ★

It was precisely this situation that Varmus, Thomson, Gearhart, and West were hoping to avoid when they testified before Congress in December 1998 endorsing government support, both financial and regulatory, of stem cell science.

For its part, WARF, in Gulbrandsen's view, was acting no differently from the way it always had—protecting the interests of the University of Wisconsin—and there was no expectation or obligation to act otherwise when it came to stem cells. "Our foundation is not set up to serve the rest of the world; it's set up to serve the University of Wisconsin–Madison," Gulbrandsen says in defending WARF's initially aggressive decisions about the cost and conditions of distributing the cells.

While WARF was making these decisions, much of the scientific community's ire inevitably fell upon Thomson. The patents, after all, had been issued in his name, and while he had assigned all rights to WARF, he became the target of his colleagues' growing frustration. "Everybody historically saw me as controlling the thing," he tells me. "Yes, there was a lot of animosity out there. And it wasn't just the patent but the distribution of the cells." He sums up the situation with his characteristic frankness. "The MTA we had sucked."

Thomson also admits that the yearly update letter took things too far. When I ask if he would have agreed to provide such a document, he scoffs. "It would have annoyed the shit out of me. Would I like to do a progress report every year as part of an MTA? Come on! I'd get pissed off."

Thomson did complain to WARF about the grievances he was receiving, but at the time, he says, "it didn't matter. It wasn't like I could do anything about it, since the major reason for it was to show our state legislature that we were keeping track of the cells."

The hue and cry, and WARF's own growing skills at juggling its various obligations—to the state, whose citizens supported the foundation's endeavors with their tax dollars, to the university, as well as to the larger community of scientists—led to some major modifications in the original MTA. And some softening of WARF's initially hard line. The price of the cells dropped from five thousand dollars for two vials to five hundred dollars, after the government negotiated that same fee for federally funded investigators. And the "reach through" clause entitling the foundation to future royalties from commercial products, which Gulbrandsen maintains wasn't a reach through but does acknowledge was written "as ambiguous as could be," was removed.

"I think WARF lost the opportunity to be the leader in making things simple and nice," says Thomson. "Over the years they figured it out, but it took forever. I think they missed their blind spot on how the academic community works. But they learned over time."

I ask Gulbrandsen if he would do anything differently, given the black eye that WARF got when it sauntered into the stem cell ring with a heightened sense of entitlement, and given the fact that he ended up agreeing with the research community on the excessiveness of some of the original terms. He pauses, then says, "People can get angry at WARF but they at least have to agree that WARF learns. We are not perfect, but we do learn. And we were trying to learn as we went."

5

A President, a Policy,
and a Predicament

August in Texas isn't known for being easy on the lungs. Even in the early hours, well before the full brunt of the viscous southern heat reaches its suffocating zenith, the first cloying waves of humidity have already begun to saturate the morning air.

August 9, 2001, was no exception. So when President George W. Bush and one of his domestic policy advisers set out for a morning jog on the chief executive's ranch in Crawford, one of the men wisely decided to say as little as possible in order to conserve his breath.

After all, Jay Lefkowitz figured, he had already done his talking. Throughout that spring and summer, he'd had dozens of private conversations with the president, sat in on numerous meetings with scientists and ethicists and legislators, and drafted a small mound of memos to brief him. The time for talking was over. Later that day, the president would sit in front of a camera in a makeshift studio in his 3,300-square-foot "Western White House" and finally let the rest of the nation in on the most anticipated decision of his eight-month-old administration. In another month, planes flying into New York's World Trade Center, the Pentagon, and a field in Shanksville, Pennsylvania, would redefine the Bush presidency and put a far more urgent domestic policy issue on the president's desk. But on that August morning in 2001, as the two men headed out into the thick Texas morning, the most heated and politically sensitive topic on the national agenda was, remarkably, stem cells.

Throughout his campaign, Bush, a devout Christian, had taken a staunchly conservative position on reproductive health issues: He was against abortion, and by extension opposed to embryo research, since studying these embryos nearly always required either harming or destroying them—an act he viewed as the taking of a life.

For three days, Lefkowitz, the president, and his trusted counselor Karen Hughes, the no-nonsense former journalist and fellow Texan who had helped to manage both his successful gubernatorial and presidential campaigns, had confined themselves to the ranch, crafting the words Bush would deliver to the nation that evening. It was the president's first televised prime-time address to the people, and the fact that the topic was not a war or a political crisis or a matter of national security may have been telling in and of itself. The decision on stem cells had taken on a life of its own in the Bush administration's first months, becoming a critical political bellwether of exactly how conservative the White House would prove to be. And a televised address, some pundits noted, had the advantage of giving Bush's decision a gravitas to reflect how seriously the president had considered his position on human embryonic stem cells—not to mention also allowing him to escape a press conference setting, in which he would have to answer questions from reporters.

The fact that a science issue—a matter of biology involving alien-sounding entities such as blastocysts, inner cell masses, and pluripotent cells—had assumed national prominence was unusual enough. But by the summer of 2001, stem cells had become much more than a question of basic biology. Wrapped up as they were in the embryos from which they came, stem cells had grown into the litmus test for the Bush administration: Conservatives were counting on the new president to reverse the stem cell policies sketched out but never executed by the Clinton administration, while liberals were holding their breath to see how much havoc the antiabortion president would wreak on a promising new field that had the misfortune of being entirely dependent on a political hot potato. Rumors had bubbled up, both in the press and among staffers, that Bush was considering banning the whole line of study altogether, snatching the rug right out from under federally funded research and forcing stem cell science to go underground, into the private sector, where practitioners were expected to police themselves and government regulations for scientific conduct and ethical behavior did not necessarily apply.

Keenly aware of the heightened interest in and sensitivity to the issue, for months Bush had intentionally kept his speech for that evening out of the hands of his usual writers; instead, he asked Lefkowitz and Hughes to come up with a first draft, which the three of them then reviewed, edited, and rewrote during those hazy August days at Crawford.

For Lefkowitz, the speech—and the president's decision—represented the culmination of an eight-month crash course in biology and ethics. "It was like going to graduate school for me," he tells me over coffee one afternoon near his law offices, in midtown Manhattan. Still boyish-looking at forty-eight, the bespectacled Lefkowitz, as befits his profession, is a good talker, and persuasive—in fact, disarmingly so. Because he's quick-thinking and accommodating, it's not immediately apparent why he has earned the nickname Viper in courtrooms. As he describes the chronology of events leading up to Bush's decision, however, it's clear he is a careful and intentional person, one who prefers to approach matters from a position of knowledge and fully informed learning. It's a trait that has earned him the respect of scientists who don't agree with his conservative position on stem cell research but nonetheless appreciate his thoroughness and open-minded view of the field. "I knew philosophically he was on the other side," says Dr. Irving Weissman, a pioneering stem cell biologist at Stanford University and one of the vocal critics of the Bush stem cell policy. "But I give him an A for credibility."

Much of that preference for learning before leaping, Lefkowitz says, came from his father, also a lawyer, who served in the Civil Service Employees Association in New York, the state's largest union. As part of his political education growing up during the cold war, his father insisted that he read three books on anti-utopias, an experience that left a deep impression on the young teen. Years later, he can still rattle off to me the titles that had such an impact on his thinking: *Brave New World*, *1984*, and *Darkness at Noon*.

Lefkowitz's relationship with the Bush family began with the elder George Bush's first presidential campaign, in 1987, when he began a career-long focus on exposing and addressing human rights violations around the world. Returning to the private sector during the Clinton years, Lefkowitz found himself back in the White House with the second George Bush as general counsel in the Office of Management and Budget. One of the first things to land on his desk when he showed up for work in January 2001 was a lawsuit. It involved stem cells.

The suit was filed by a California group known as Nightlight Christian Adoptions, which took issue with the fact that the NIH, acting on the recommendations of the Human Embryo Research Panel, was accepting proposals for research grants on human embryonic stem cells (but not, thanks to President Clinton's ban, on studies with embryos created solely for research purposes). Nightlight is an adoption organization that among its other services, matches excess IVF embryos with couples willing to adopt them and raise the resulting children as their own. The suit charged that if the NIH were to approve grants in the field, it would be breaking the law, as laid out by the Dickey-Wicker amendment. It demanded that NIH stop all stem cell activities immediately.

Whatever its legal merits, the lawsuit gave the Bush White House a good reason to look at the stem cell policies that had been set in motion during the Clinton era. And it provided the conservative-leaning new president an opportunity to take on the issue and make it his own.

Technically speaking, Bush didn't have to do anything about stem cells. By the time he entered the White House in January 2001, the NIH had already received a handful of requests for grants to work with IVF-based embryonic stem cells. Scientists had until March 15 to apply for the federal funds, at which time the agency would begin reviewing them and writing checks for those they deemed most worthy. Harold Varmus had also ordered up a set of guidelines evaluating incoming research proposals. Everything was in place for the agency to fund, for the first time, studies on human embryonic stem cells.

Or so it seemed. Having campaigned on his strong conviction that embryos should not be destroyed for any purpose—even for research that might benefit those suffering from incurable diseases, and even if those embryos were excess embryos that parents and IVF clinics had decided to discard—Bush felt that the Clinton policy didn't square with his own administration's beliefs. While he never specifically addressed embryonic stem cells during the campaign, as a candidate Bush had responded with the following statement to a questionnaire by a Catholic group on the issue of embryo research: "Taxpayer funds should not underwrite research that involves the destruction of live human embryos."

So rather than quietly quashing the NIH program, Bush officials instead may have sensed an opening—a chance to establish their candidate turned chief executive with a weighty decision on a thorny, controversial issue that could showcase the new president's thoughtful and conscientious side. "A lot of the conservative groups saw this even then a little bit as a surrogate for abortion," says Lefkowitz, zeroing in on the most potent argument against the field, regardless of how valid the position may have been. "They reacted very strongly and there was one of these little grass-fire storms."

Within months of Bush taking office, the fire was gaining momentum, and the White House felt compelled to take action. So the education effort began. Stem cell scientists, bioethicists, religious leaders, patient advocates, representatives from Congress, cabinet members—all traipsed through the Oval Office during the spring and summer months as Bush wrestled with the question of whether taxpayer dollars should be spent on embryonic stem cell studies. On the one hand, his firm conviction that life begins at conception, and that no life should be destroyed to serve another, made the field, however promising, ethically troublesome. Could anything justify the taking of these lives, no matter how undeveloped or unformed they were? On the other hand, millions of Americans, to whom he was also accountable, suffered daily from untreatable diseases such as diabetes, Parkinson's, Alzheimer's, and spinal cord injury—all conditions that could someday be cured with stem cell therapies. Was it ethical to deny these citizens a chance for a cure?

Cynics saw the ultimate spin campaign swirling out of the West Wing, as subtle as frantic smoke signals spiraling in the distance. The message? That the president was intentionally taking his time, devoting deep thought to resolving, in King Solomon-esque fashion, this increasingly complex scientific and moral conundrum. Why, these skeptics wondered, was Bush's position suddenly so angst-provoking? Bush, the Republican nominee, had seemed sure of his position against embryo research. What had changed, these pundits wondered, to make that conviction waver? The science? The ethics?

What changed, essentially, was the pressure. Pressure from the liberal-leaning, pro-abortion left, which was testing the limits of the new president's well-known conservative views, as well as pressure from the antiabortion right, which had voted Bush into the presidency and was eager for a show of executive force for their views. "Various conservative and pro-life organizations were agitated," Lefkowitz says of the growing expectations that the Bush administration deliver on its campaign promise of conservative values. "They were challenging the new administration, if you will. They wanted the new administration to prove itself."

Complicating matters was the growing divide within the president's own party. A contingent of staunch antiabortion Republican leaders in Congress, led by Senator Orrin Hatch, a Mormon from Utah, were voicing a strong affinity for stem cell research, even though embryos would be destroyed in the process. After studying the issue for more than two years, Hatch concluded that a pro-life position allowed for stem cell research on embryos that were destined to be discarded, and that a several-day-old embryo in a dish did not have the same moral stature as one implanted into the womb, or a five-year-old child. For other representatives, support of the science came from a more personal place—Senator Strom Thurmond of South Carolina, who had to watch his daughter struggle with type 1 diabetes, was persuaded that NIH's involvement in stem cell research would allow for faster and safer cures, while Senator Gordon Smith, a Republican from Oregon, decided to support stem cells in honor of family members who had been debilitated by Parkinson's. Strange bedfellows, indeed. But it was, in the words of some Republican strategists, the new compassionate conservatism.

Within the administration itself, staff members were also aligning themselves on one side or the other. Karl Rove, true to his duties as a political strategist, strongly opposed federal funding of the science, concerned that the administration's support of stem cells would not be viewed kindly by the president's strongest conservative supporters—supporters he would need not only on Capitol Hill, but in the next election. Tommy Thompson, the newly appointed secretary of Health and Human Services (HHS) and former governor of Wisconsin, who had proudly praised James Thomson's groundbreaking work—at the University of Wisconsin—just three years before, urged the president to support the studies.

In fact, it was Thompson, Lefkowitz tells me, who in part prompted the White House to push stem cell research to the top of Bush's agenda. Testifying at a Senate hearing in March just after his appointment to HHS, the straight-talking Midwesterner's response to a question about the Dickey-Wicker amendment must have caused the new administration to suck in its breath: "There is a law that has been passed by Congress that prohibits the use of federal dollars for stem cell research that comes from a destroyed embryo. . . . I am troubled, as you probably are, by the law."

While Thompson was referring to his own misgivings, and not those of the White House per se, as a member of the cabinet he was expected to reflect the views of the administration, so not surprisingly, the next day White House reporters immediately questioned press secretary Ari Fleischer about the president's position—had it changed from the campaign? Was he inclined to be more supportive of the research than many had assumed? Unable to answer, Fleischer came to Lefkowitz for help. "He came down to my office and said, 'Jay, what do we say? Do we have a position on this issue?'"

The scientists, in the meantime, watched in helpless frustration as politics pulled their studies deeper and deeper into a bureaucratic morass. After seeming to be on the right track with Varmus's decision to draw up guidelines and invite applications for grants, just three months into the new administration the research again fell prey to the political tug-of-war, when Tommy Thompson ordered a moratorium on NIH's plans to fund the first studies on human embryonic stem cells until the agency could complete a review of the field. He assigned the task to the Office of Science and Technology Policy (OSTP), an HHS office that was responsible for all policy issues regarding research conducted by NIH. Led by Lana Skirboll, a pharmacologist whom Varmus appointed, its staff is made up not of scientists but of government officials specializing in navigating the ins and outs of federal policy-making and budget allocation.

The moratorium meant that the NIH's planned meeting to review the first grant proposals had to be scuttled. Two weeks before the review committee was set to evaluate these submissions, the dozen or so members serving on the committee were called, one by one, and told their services would no longer be needed. Most were not surprised in the least to get the news.

★ ★ ★

For the scientists, the cancellation of that meeting was an ominous sign. Until then, stem cell advocates and scientific experts were keeping a low profile, hoping that the existing stem cell policy would sneak past the new occupant in the White House, and banking that once the studies got under way, the merits and potential of the science would speak for themselves. Their hope, which proved unfounded, was that Bush would leave things be.

Taking their cue from the Nightlight suit, a group of well-heeled leaders in the business field—all patients with intimate knowledge of the challenges of living with diseases that could be treated with stem cells—initiated their own legal action against the government. The cancellation of the review meeting provided them with a good reason: It represented, in their view, an unacceptable interference with an established government policy. Fearing a dragged-out process in which federal funding of stem cell studies would continue to languish while tangled up in bureaucratic politics, the savvy attorneys also enlisted the power forwards from the scientific side—Jamie Thomson, Doug Melton, and leading Australian experts Alan Trounson and Martin Pera. For added luminary impact, Christopher Reeve also joined the suit.

What the scientists were fighting for was simply the status quo. The fact that NIH was calling for stem cell proposals was thanks in large part to Harold Varmus, who had found a clever way around the Dickey-Wicker restriction. Soon after the publication of Thomson's and Gearhart's papers, he wrote to the legal counsel at the Department of Health and Human Services, to which NIH reports, and asked for advice. Would study of embryonic *stem cells* be considered research on an embryo, he asked, and if not, then wouldn't they be fair game for government-funded scientists?

Harriet Rabb, now at the Rockefeller University, was the general counsel for the department at the time, and after a critical review of how the embryo was defined in the Dickey-Wicker amendment, ended up making an important decision that even Lefkowitz agreed with. She determined that an embryo is an organism when it is implanted in the uterus and allowed to develop, and that by definition the stem cells, which were never intended to be transferred to a woman for further development, are not embryos. And that they were therefore eligible for study with taxpayer dollars.

"Ultimately I basically agreed that as a legal matter, we had the right to use federal funds for stem cell research," Lefkowitz tells me. Then he quickly adds, "But we were not required to, so therefore the president had to make an executive decision about whether we should."

The lawsuit that greeted Lefkowitz on his first days on the job maintained otherwise, contending that the Rabb decision, and therefore NIH funding of embryonic stem cell research, would be against the law. But Lefkowitz made a legal determination not to fight the policy on legal grounds. That choice would prove critical in shaping the Bush administration's ultimate decision on what to do about stem cells.

The Rabb decision was a brilliant preemptive strike on Varmus's part. While he describes it as a no-brainer, he acknowledges that he "needed to have a good legal statement. I needed something. People today still don't think it's legally grounded, though I disagree. We had to say that stem cells are not embryos, and that the prohibition does not apply."

Already, Varmus could feel the foundations that had been set during the Clinton administration starting to sink into the political muck. Following Rabb's decision, he emphatically informed Congress, in several appearances on the Hill, that he intended to go ahead with the funding of human embryonic stem cell research. "No federal funds were used, no laws were broken in producing these stem cells, and we have determined that no laws would be broken if federal funds are used to support work with them once they have been derived," he testified.

But as soon as the Bush administration suspended all of NIH's efforts in the field, it was as if all of Varmus's efforts had never happened. The NIH was back where it had started.

★ ★ ★

When I ask Lefkowitz why the president felt the need to scrap existing policy on stem cells and start from scratch, he intimates that the reasons were driven more by political than by scientific expediency. "Bush felt he had a responsibility, an obligation to determine how consistent with federal law federal funds would be spent," he says.

And thus began that march of the stem cell brigade—the near constant stream of experts and advisers from a broad range of different fields who came, at the request of the White House, to educate and inform the president on the subject throughout the summer months.

Doug Melton was one of those called, and his experiences provide a revealing peek into the dynamics within the White House. He went not as the codirector of the Harvard Stem Cell Institute, but as a member of the Juvenile Diabetes Research Foundation, to share with the president his reasons for believing in the potential of stem cells for advancing diabetes research. Given his prominence in the field, Melton ended up talking to Vice President Dick Cheney and Karl Rove, as well as the president. Making the trip at his own expense, he was eager for the chance to explain why he felt the field was so critical for the future of medicine, and why he personally felt the clock was ticking on finding a cure for diabetes.

His conversation with Cheney revolved around the promise of stem cells in treating heart disease, an especially relevant topic as the vice president had recently had a cardiovascular procedure. "Mr. Cheney has a very quick and curious mind," Melton tells me during one of our talks in his Cambridge office. "I remember teasing him that he was so good he could come give lectures on heart stem cells. He asked me excellent questions about the upsides, the downsides, and the ethical issues."

While Cheney seemed open to Melton's arguments, in the end, he told the scientist, the decision was the president's to make.

But before he could meet Bush, Melton was also ushered into Karl Rove's office. At the time, Rove wasn't the household name he would soon become, and wasn't yet established in the public's mind as the mastermind of the Bush White House. Melton confesses that he didn't know who he was, and found particularly odd Rove's agenda for the meeting. "He tried to convince me that what I was saying was scientifically not right," he says, "and that even if I, meaning the group of scientists working on stem cells, were able to make stem cells into tissues or body parts, the immune system would reject them. The immune system is not something that has escaped our attention," he adds with a laugh. "I remember thinking that if he didn't want the research to go forward, he should be telling me about other legislative reasons, political reasons, things I didn't know anything about. It seemed odd that he tried to argue me out of a scientific point."

That, coupled with the way Rove's assistant constantly interrupted the discussion with a flurry of quietly delivered pink notes requiring Rove to at one point get up and leave the room, left Melton frustrated with the visit. "I was disappointed that he didn't show me the courtesy of actually setting aside twenty minutes to listen to what I had to say. It was like, 'I don't know or care who you are; someone asked me to have this meeting and I'm busy with other important things,'" says Melton. "That's the sort of impression I left that meeting with."

The discussion with the president, says Melton, was only slightly better. Bush's opening remark, Melton recalls, actually startled him. "He sat down, asked me to sit next to him, and said, 'Here's your chance to pitch the President of the United States.'" Melton, not one to be distracted or easily ruffled, was caught off guard. "I remember thinking at the time, first, this is surprising—that he's talking about himself in the third person—and second, that I wasn't really there to pitch anything."

He may, however, have actually been picking up on a subtle but important clue about how the president and his top aides were approaching the stem cell issue. Speaking at an appearance in London in July, Bush himself noted that "This is way beyond politics. The American people deserve a president who will listen to people and make a serious, thoughtful judgment on this complex issue, and that's exactly how I'm going to handle it."

But some political analysts weren't buying the rhetoric, speculating that the president had already made up his mind to restrict funding of stem cell research, and was going through the motions of deliberation and thoughtful analysis for political show. Melton, coming to the White House as a representative of JDRF, would certainly be making the case for more expansive funding, and that was the way Bush was viewing the visit—as a pitch for him to change his mind.

What bothered Melton most about the president's remark was its apparent lack of gravity. For Melton, stem cell science had become much more than an abstract debate between believers and nonbelievers. For him, it represented the very real possibility of a much better alternative to the daily injections that his son endured several times every day in order to stay alive. His role, as he saw it, wasn't to sell one position or another. It was to stand up for science. "I was there to explain why I thought stem cell science and embryonic stem cell science in particular was an exciting new door to how our bodies work, and equally important, a tool or reagent to lead to cures for terrible diseases for which we don't have any treatments," he says. "Diseases that are costing the nation huge amounts of money, not to mention what I personally feel is more important, which is the pain and suffering of all these children."

Melton's "pitch," as it were, acknowledged that certainly stem cell science was ethically challenging, but he made the case that, at least in his view, the ethics seemed straightforward. The embryos discarded by IVF clinics, he told Bush, were destined to be destroyed anyway, and if they could be saved, they would be a valuable resource for scientists to study. And the knowledge gleaned from embryonic stem cells, which could someday lead to treatments that would help millions of Americans, would certainly present a better choice than simply throwing the embryos away.

At one point, Melton even tried to appeal to Bush's sense of legacy, informing him that his support of stem cells could be his "moon shot." The president's reaction, says Melton, was "pretty flat." Walking out of the meeting, he felt he hadn't done much to inform the president, much less convince him of anything.

Bush continued to have similar discussions with experts throughout the summer of 2001. But in an odd effort to keep the president's deliberations under wraps, Lefkowitz says, many of them were not noted formally on Bush's calendar. Instead, thirty-minute blocks would be labeled as "personal time" or attributed to key advisers, such as Chief of Staff Andy Card, during which scientists, ethicists, and advocacy organization representatives would wrestle with the complexities of stem cell science.

As interest in his position on stem cells escalated, Bush coyly avoided reporters' questions on when he would make his decision, and what that decision would be. Administration officials leaked only hints of where the president was heading; Bush himself only flashed his hand in a June letter to the Culture of Life Foundation, an organization formed according to the principles of the Catholic Evangelium Vitae, or Gospel of Life. "I oppose federal funding for stem cell research that involves destroying human embryos," he wrote. "I support innovative medical research on life-threatening and debilitating disease, including promising research on stem cells from adult tissue."

Adult stem cells had become the foundation of the pro-life movement against ES cell research. Why support the killing of embryos, these groups argued, when adult stem cells, obtained without the need for embryos, could prove just as scientifically and medically useful? Open questions about how relevant they were to finding cures for disease, and how equivalent they were to their embryonic counterparts, made them questionable stand-ins in most scientists' view, but for those opposed to destroying embryos for research, they represented a logical solution to the stem cell debate.

That was the case for Richard Doerflinger, who oversees pro-life activities at the United States Conference of Catholic Bishops and became a vocal critic of embryonic stem cell studies. An effective speaker with a graying beard and avuncular demeanor, Doerflinger had been one of the few pro-life representatives attending the public meetings of the Human Embryo Research Panel in 1994, and, as they did for many of the panel members themselves, those discussions helped to shape his own views on the morality of using embryos for stem cell studies. While the scientists and many bioethicists were arguing from a utilitarian perspective that it was better to conduct potentially lifesaving research on IVF embryos that would be discarded, Doerflinger and the pro-life movement, including the Family Research Council and National Right to Life, were attempting to make a more philosophical assertion. "Catholic morality is largely based on the understanding that the first result of acts is the result they have on us, the people who do them," he says. "They have an effect on our character. So we are involving ourselves in taking this step against life in a new way. It's not just about 'Is this embryo going to die anyway?'"

Doerflinger himself admits to coming to this perspective via personal experience. As a teen, his older brother was involved in a car accident that left him in a coma for several months. Told by the doctors that he would likely not wake up, Doerflinger's family agonized over whether to keep his brother on life support. As they struggled with the decision, his brother regained consciousness, and while confined to a wheelchair, continues to enjoy a full life. It was nearly impossible after that experience, says Doerflinger, who had aspirations of becoming a doctor, to support any reason for taking a life. "We can never take a life, an innocent life, to help another life," he tells me.

But are adult stem cells as potent as their embryonic predecessors? Without denying the utility of adult cells in treating disease, scientists argued that in order to truly harness the power of stem cells, they should be allowed to unlock the developmental potential of embryonic as well as adult versions of the cells. But the political rhetoric was forcing an artificial divide between the two, as if the government had to choose one or the other.

In an attempt to break down that barrier, HHS Secretary Thompson commissioned the NIH in spring 2001 to conduct a thorough review of existing knowledge of both embryonic and adult stem cells. The report was the most comprehensive analysis of the science on stem cells to that point, and listed an impressive roster of eighty-three leading experts who were interviewed during the two-month process. Ariff Bongso, the Singaporean scientist who isolated the first human embryonic stem cells but failed to keep them alive, was consulted, as were experts in Israel, Japan, Australia, and the United States. They, along with the NIH authors, combed through dozens of scientific papers in the field. "To date," the report concluded, "it is impossible to predict which stem cells—those derived from the embryo, the fetus, or the adult—or which methods for manipulating the cells, will best meet the needs of basic research and clinical applications. The answer clearly lies in conducting more research."

For stem cell scientists, it was hardly astounding or headline-grabbing stuff—it was simply a restatement of the argument they had been making all along. But the report was tailor-made for the nonexperts who now found themselves in the position of making critical decisions about the science. The report contained color diagrams of the process required to isolate and grow up embryonic stem cells; it touched on the latest science in how stem cells might help to yield new treatments in diabetes, neurological disorders, and heart disease. But perhaps most important, it laid out, for the first time in one place, the benefits and disadvantages of embryonic stem cells and adult stem cells—based solely on the science. Thompson had specifically asked the group not to weigh the ethical considerations, because other panels, including the Human Embryo Research Panel in 1994 and the National Bioethics Advisory Commission, had already attempted to wrestle with that aspect.

Adult stem cells, the authors wrote, were more difficult to obtain from patients since they were rarer. And once isolated, they tended to be more finicky on the culture plates; it was hard to grow adult stem cells in sufficient quantities to eventually transplant them, say, into an ailing heart or a diabetic's pancreas. And that was assuming that the cells could be coaxed into developing into the specific cell types that a patient would need.

Embryonic stem cells, on the other hand, possessed remarkable plasticity and, in stem cell lines established thus far, appeared to grow more happily and robustly in the lab. The cost of this pluripotency, however, was the tendency to form tumors known as teratomas, the otherworldly mixture of a variety of differentiated cells. But according to the experts consulted, that danger could be averted with careful and proper culturing of the stem cells to keep them in their controlled, pluripotent state.

Rather than resolve differences between those for and those against embryonic stem cell research, however, the report only galvanized those on either side of the debate. On Capitol Hill, senators, representatives, and experts interpreted the report in their own pro or con ways. Testifying at the Senate hearing at which the report was released, Senator Orrin Hatch asked, "Why shouldn't these embryos slated for destruction be used for the good of mankind? . . . To me, the morality of the situation dictates that these embryos, which are routinely discarded, be used to improve and extend life. I do not believe it would be wise to cut off support for embryonic stem cell research."

Stepping to the microphone shortly after Hatch, Doerflinger reiterated his group's position that funding such studies would be illegal, immoral, and unnecessary. Citing growing studies on the promise of adult stem cells in treating disease, Doerflinger took particular umbrage at the ethical ease with which some condoned research on excess IVF embryos that would be destroyed anyway. Calling such reasoning "sophistry," he testified, "We do not kill terminally ill patients for their organs, although they will die soon anyway, or even harvest vital organs from death row prisoners, although they will be put to death soon anyway."

The pressure was mounting—from both sides—for President Bush to take a position. Weeks before the report was released, Senator Hatch had penned a twelve-page letter to the president, addressing the apparent contradiction of his antiabortion position and support of stem cells head-on. "Stem cell research facilitates life," he wrote. "Abortion destroys life; this is about saving lives." It was possible, Hatch urged Bush, to be both pro-life and to support the study of embryonic stem cells originating from excess IVF embryos. Relying on a position that many bioethicists were beginning to take—that the moral status of an embryo is not equal across its entire development—Hatch told reporters, "I just cannot equate a child living in the womb, with moving toes and fingers and a beating heart, with an embryo in a freezer."

Thirty-eight House Republicans wrote their own appeal to Bush, countered by an equally intensive effort by anti–stem cell House members urging the president to withhold federal support for the field. The party's House leaders—Richard Armey, the majority leader; Tom DeLay, the majority whip; and J. C. Watts, the Republican conference chairman—released a statement questioning the moral soundness of the Rabb decision. "The federal government cannot morally look the other way with respect to the destruction of human embryos," they said, "then accept and pay for extracted stem cells for the purpose of medical research."

And it wasn't as if the scientists were standing by idly, either. For academic researchers, dependent as they were on government funding through NIH to support their studies, no new work was planned with embryonic cells. Universities that relied on federal grants to support an entire campus's worth of careers and experiments could ill afford to allow their one or two stem cell scientists to push on in defiance of the moratorium and put an entire facility in jeopardy of losing its government dollars.

Private researchers, on the other hand, could only see green—green lights to proceed where their ivory tower colleagues could not, and a potential financial windfall if their efforts panned out. At the very least, if they could successfully and reliably generate embryonic stem cells, they would hold the key to a sought-after resource—stem cell lines—if government regulations opened up to welcome taxpayer-funded studies. As it turned out, two private facilities took advantage of the regulatory limbo and did exactly what legislators and pro-life groups had worried they would do: They created human embryos specifically for research purposes, without any government or public oversight.

The first attempt came from a fertility clinic in early July. Paying women to provide eggs, and men to give sperm, scientists at the Jones Institute for Reproductive Medicine, in Norfolk, Virginia, mixed the two gametes together in a dish and allowed the fertilized embryos to grow for four to five days, at which point they scraped off the embryonic stem cells bundled inside the hollow blastocysts. From 162 fertilized eggs, only fifty embryos began to divide and grow, forty of which yielded three successful stem cell lines. (There was a certain irony that these pioneering research embryos were generated at Jones. It was a team of physicians at the Virginia clinic that had created the first American test tube baby, in 1981; while the skill set of the doctors there certainly made them naturals for manipulating embryos for stem cells, none of those early physicians could have imagined they would be working so assiduously to create embryos that someday would not help a woman to become pregnant at all.)

Not to be outdone, researchers at Advanced Cell Technology (ACT), in Marlborough, Massachusetts, now led by Michael West, who had by this time been forced out of Geron, revealed a day later that it had been working for a year on generating the first human clone. The project was not, stressed West, an effort to make clones of people, but to facilitate a method for obtaining patient-specific stem cells. Using the same technique that Ian Wilmut had used in cloning Dolly, West's team was hoping to clone adult skin cells and grow the resulting embryos just long enough to extract their stem cells from them. Aware of the moral uneasiness that the work engendered, West reported that the days-old embryos had their own bodyguards, as well as constant video surveillance, to

ensure that they would not be implanted into a woman's womb to generate the first human clone.

Neither company, however, had achieved resounding success; Jones's
foray had resulted in a less than 0.2 percent success rate in culturing a stem
cell line, while ACT had yet to see its first cloned embryo live long enough
to yield the cells. But the impact of their announcements was clear—the
science was here, and if the government wouldn't fund or oversee it, then
there were plenty of privately supported companies that would.

To proponents, these advances were a warning shot, a glimpse of the
wild west culture to come if private enterprises, guided by good intentions
and financial mandates, hijacked control of the technology. "This type of
research is indicative of the problems we will continue to encounter if
we don't allow federal funding with strict research guidelines for embryonic stem cell research," Hatch warned. "Without stringent, NIH ethical
requirements, we are opening the door to an array of different research
standards, which I believe could create some serious consequences."

For Bush, the stunning announcements only meant that the chorus
was getting louder. "People started calling in, and members of Congress
wanted to be heard on the issue," says Lefkowitz. "But we kept it very,
very quiet. Even though everybody knew the president was thinking
about this issue, it was a very small group of people—four to five people
in the White House—who were actually involved in discussions with
the president. It was really just Karl, Karen, Andy, and myself."

★ ★ ★

"Which one of my children would you kill?" John Borden asked members of a packed House as his wife held up an image of three bundles of
cells. Two of those bundles were now sitting in their parents' arms, nine
months old and wiggling under the scrutiny of their theatrical congressional debut. "Which one would you take?"

As a new father, Borden was testifying to prevent the government
from funding research on embryos. One of the embryos in the picture,
Borden explained, did not survive. But all three were destined to die until

John and his wife, Lucinda, decided to adopt them from another couple that had frozen them and no longer needed them. Luke and Mark Borden were the human face that conservatives had been desperately looking for in their fight to ban embryonic stem cell research. While their opponents paraded a seemingly endless line of patients suffering from diabetes, spinal cord injuries, Parkinson's, heart disease, and other conditions—along with an increasingly visible celebrity contingent, with the likes of Christopher Reeve and Nancy Reagan, whose husband, the former president, suffered from Alzheimer's—stem cell critics had as their poster child only a faceless, nearly invisible embryo that barely filled the tip of a pin.

But now, thanks to Nightlight Christian Adoptions, those opposed to stem cell research also had a face for its cause. Several very adorable faces, as it turned out, of which the Borden babies were only the vanguard.

All the lobbying from both sides only heightened the intensity of Lefkowitz's job, which was to keep President Bush informed of the science and ethics of stem cell science. Drawing upon scientific as well as literary resources to prepare the president, at one point, recalling a passage from *Brave New World* that would make his point about the hubris in controlling reproduction, Lefkowitz even brought his copy of the novel into the Oval Office for a dramatic reading. He found himself driving to work earlier and earlier to keep up with the president's barrage of questions and requests for more information. A notorious advocate of the early to bed, early to rise philosophy, Bush, Lefkowitz admits to me with a smile, started his calls quite early.

"He would sometimes call me early in the morning to follow up on something he had read overnight in a memo I had given him, or a conversation he had had," Lefkowitz says. "I didn't want to get calls from him at seven o'clock in the morning. I was still coming out of the shower in Chevy Chase, Maryland, and he'd say, 'How soon can you be here?' I'd say, 'Twenty minutes,' and I could hear the dejection in his voice because by then he'd be going on to something else already."

By mid-July, the press was percolating with rumors of a compromise position rather than an outright ban. Bush, some officials speculated, was toying with allowing federal funding of only existing stem cell lines. For scientists, however, it was less of a compromise and more of a concession that would limit their ability to create, study, and compare as many ES cell lines as possible. Buttressing this position, a report by MIT's Rudolph Jaenisch detailed a startling range of genetic variability in mouse embryonic stem cell lines—and mouse clones grown up from the cells. Different lines had different tendencies to stay undifferentiated and, presumably, would generate slightly different types of specialized cells if they were allowed to develop fully. The science was so new, Jaenisch argued, that it didn't make sense to limit study to just a few cell lines.

Outside the White House, proponents and opponents continued their pressure on the "agonizing" decision facing Bush. Nancy Reagan, having watched for almost a decade the devastating effects of Alzheimer's disease on her husband, an icon of the antiabortion, pro-life conservative view, wrote a letter to Bush urging him to support stem cell studies. Michael Deaver, a Reagan loyalist and friend of Bush, hand-delivered the one-page letter to the White House.

Even the Pope weighed in, albeit indirectly, during a visit with Bush in July, noting in his formal statement to the president, "A free and virtuous society, which America aspires to be, must reject practices that devalue and violate human life at any stage from conception until natural death."

As the stem cell summer of 2001 wound its way into the dog days of August, members of the House on opposite sides of the issue laid down one more ultimatum. With 260 members voicing their support of federal funding of ES cell research, opponents of the science introduced legislation that would ban human cloning.

Cloning had been a long-standing thorn in the side of the stem cell debate. Critics of ES cell research opposed any form of human cloning, including cloning that would create only days-old blastocysts never intended to become a human being but instead provide a source of human stem cells. Proponents of ES cell research urged a more refined view of the technology—most, including the leading scientists in the field such as Melton, Thomson, and Wilmut, opposed cloning for reproductive purposes—that is, cloning an embryo to implant into a woman's womb so that it could develop into a baby. What they advocated instead was therapeutic cloning, which allowed cloned blastocysts to survive to five days, only long enough to yield embryonic stem cells.

By raising the cloning issue, however, those against stem cell research highlighted the slippery moral slope that they predicted would lead from stem cell studies to embryo farming and baby-making clinics. With such dangers front and center in the debates over the cloning bill, these critics hoped to convince more members of Congress—and the public—of the hazards of the science. The outcome might also influence the president as he continued to struggle with his own decision about what to do about stem cells. "There are certainly people who are not with us on the stem cell issue who are with us on cloning," noted Doerflinger in a *New York Times* interview.

The strategy worked. With a 265 to 162 vote, the more restrictive bill passed, outlawing all forms of human cloning and carrying a punishment of up to ten years in prison. In another ominous sign for stem cell advocates, Bush expressed his support of the bill. The legislation, he said, "is a strong ethical statement, which I commend. We must advance the promise and cause of science, but must do so in a way that honors and respects life."

At the White House, in an effort to forge through the ever-thickening tangle of moral and political concerns ensnaring the administration, Lefkowitz recalls relying on a useful teaching tool bioethicists and lawyers often use. "I remember a discussion with [the president] on the moral conundrum of what we could do if we had a death row inmate who was seconds away from being executed by lethal injection, and there was no question about the morality of taking that person's life—the state and society had already determined that the person forfeited the right to life. Yet even in the five seconds or thirty seconds before injection, we wouldn't countenance harvesting his organs, even though those organs might save people's lives. Now, I realize that's not a perfect parallel," Lefkowitz says. "But these are the kinds of issues we raised."

Three days later, Bush, Lefkowitz, and Hughes left for Crawford.

6

Fighting Back

On vacation in Woods Hole, Massachusetts, Doug Melton listened to President Bush's speech with his son, Sam, and his daughter, Emma. Without a television, the family heard the address on the radio, and as soon as it was over, Melton was on the phone with representatives from the Juvenile Diabetes Research Foundation, which had been actively lobbying the president to support embryonic stem cell research.

Bush's speech was a masterful demonstration of the classic cliffhanger. Written largely by Hughes and Lefkowitz, with Bush weighing in on drafts, it was a verbal juggling act. As Bush played with the nation's expectations, tossing the pros and cons of the debate from one hand to the other throughout the eleven-minute announcement, scientists and lawmakers alike were hard pressed to figure out in which camp the president would finally find himself. "Based on preliminary work that has been privately funded, scientists believe further research using stem cells offers great promise that could help improve the lives of those who suffer from many terrible diseases," Bush noted, before countering with the observation that "Research on embryonic stem cells raises profound ethical questions, because extracting the stem cell destroys the embryo, and thus destroys its potential for life. Like a snowflake, each of these embryos is unique, with the unique genetic potential of an individual human being."

It wasn't until the last minutes of the address that Bush's decision finally became obvious. "I have concluded that we should allow federal funds to be used for research on these existing stem cell lines, where the life and death decision has already been made."

The sigh of relief was almost audible—from scientists, from patients, and from the stem cell advocates who had begun to view the primitive cells as the holy grail in the search for a cure. He had, after all, given the NIH the green light to proceed with funding embryonic stem cell studies. "We were supportive of the president's decision," says Peter Van Etten, then president of the JDRF, which initially counted the decision as a victory because studies of the cells were not banned altogether.

Melton, too, was pleasantly surprised. But he couldn't get over the statement Bush had made about the number of human embryonic stem cell lines—more than sixty—that he said existed and, by implication, were ready and available for immediate study.

To the public, Bush's decision appeared as Solomonic as the president himself did that evening, serious and reasoned as he outlined the logic behind his decision. To stem cell scientists in the know, however, there was a relatively rapid rebound from the initial relief. Where were these lines? How had the government found them? And how many were ready for distribution and immediate study?

Melton was desperately hoping that the president was right. As a father of a child with diabetes, he was certainly counting on the existence of those lines. But as a scientist, he had to temper parental expectation with the cold reality of the facts, something his brain realized even if his heart was reluctant to admit. As far as Melton knew, the human embryonic stem cell lines that were stable and of good enough quality to use for meaningful research were few and far between. Jamie Thomson's lab had certainly generated a handful, as had Roger Pedersen in San Francisco and a few others overseas. Of the lines that he was aware of, Melton also knew that some were difficult to obtain—the U.S. lines were patented and came with what he and Harvard's lawyers viewed as an onerous licensing agreement—while others were still in their own embryonic stages, not stable or mature enough to be divided up and distributed outside of the labs where they were generated. To hear Bush announcing the existence of a solid set of more than sixty stem cell lines was both enticing and ominous to Melton. If it were true that such lines would be ready for research, fine. But knowing the still-small world of stem cell science as he did, he suspected that the actual number was much smaller. And if *that* were true, then the science he so desperately wanted to conduct had already been taken hostage by politics.

As more researchers voiced their skepticism—in the press, to each other, and to the NIH—it became obvious that the administration had prematurely seized upon the first number provided by the NIH's Office of Science and Technology Policy. Just days before Bush left for Crawford in August, the office's director, Lana Skirboll; Ron McKay, a stem cell scientist with the National Institute of Neurological Disorders and Stroke; and Allen Spiegel, then director of the National Institute of Diabetes and Digestive and Kidney Diseases, visited the White House with a list. On it, they told the president, was every human embryonic stem cell line in existence around the world. In total, they numbered more than sixty.

How they got to that figure became a relentlessly disputed topic that kept journalists and government officials at the Department of Health and Human Services and NIH occupied for the next several months. The day after the president's address, the questioning began. Melton told reporters, "The president is privy to information unknown to me and the scientific community."

Indeed, Skirboll and Secretary Thompson responded, the president, the Department of Health and Human Services, and the NIH did have knowledge of additional stem cell lines. Because the creation of human embryonic stem cell lines had not been sanctioned by the U.S. government, all of the American lines were generated behind proprietary closed doors that academics would not have access to. It turned out Skirboll had sent her staff on a countrywide search for these hidden depots of cells, as well as the lines generated overseas.

Roger Pedersen had been the target of one of these hunts, although he wasn't aware of it at the time. By that time, Pedersen had made two human embryonic stem cell lines, using the same method Thomson had, from extra IVF embryos. One day in early 2001, he was visited in his Parnassus Street lab at the University of California, San Francisco, by a gentleman from the NIH's Office of Science and Technology Policy. He asked a series of probing questions about Pedersen's work, and while he never directly made clear what his purpose was, Pedersen naturally suspected that the official wanted to ensure that no federal funds, however indirectly, had contributed to the creation of those cells. "I inferred it by the sort of interrogative approach he took," Pedersen tells me. While Pedersen had been careful to use only Geron money, supplemented by some privately raised funds from the university, he was nervous that the Bush administration was going to run strict audits of studies like his in order to ensure that not a single penny of taxpayer dollars had been used to pay for even incidental costs, such as electricity, equipment, or even housekeeping services to keep the labs clean.

Startled, and frightened that his studies could put the entire UCSF bundle of NIH grants at risk, Pedersen says that soon after the visit, they stopped. "We stopped doing derivations of embryonic stem cells, and we started establishing an off-campus site for that activity."

In retrospect, however, he now believes the true purpose of the review was to confirm how many stem cell lines Pedersen had actually derived, and what kind of consent the university had required from the embryo donors, in order to include them in the president's tally.

What Skirboll and her team didn't make as clear during their visit with the president about their tally—or, according to other government officials at the Institutes, what the White House didn't really want to hear—was that not all of the lines were ready, robust, or even real in the sense that they were happily growing in an incubator in some researcher's lab and ready to be studied. Some of the lines were only days old and had not yet been characterized, meaning they were still adjusting to being removed from their blastocysts and relocated to their new homes on petri dishes, so it wasn't clear yet whether they would survive and be sustainable. Others were tied up in licensing and patent tangles that would make it nearly impossible for U.S. scientists to obtain them, while still others were simply too weak to endure the passaging and dividing that would be required to distribute them to interested researchers.

The day after the president's address, Health and Human Services Secretary Thompson, the field's staunchest supporter in the administration, immediately defended the sixty-odd lines with a confidence that he was forced to dial back in the coming weeks. "They're diverse, they're robust, they're viable for research," he responded emphatically at a press conference to the skepticism of experts in the field.

As the back-and-forth between the administration and the scientists over the number of lines ready for research continued over the next several months, Melton quickly tired over the debate. The number, he knew, was simply a smoke screen for a far greater problem. "I was surprised on August ninth," he tells me. "But I was then truly alarmed, disappointed, and a little angry when I discovered that the NIH had agreed to all those stupid numbers for political reasons knowing real well that some of them did not exist, would never be available, and were a grotesque exaggeration for political reasons."

The government, despite Bush's limited approval of embryonic stem cell studies, was clearly reticent about diving into the field with as much enthusiasm as it had greeted sequencing the human genome, or even gene therapy. Because of his work on mouse embryonic stem cells, Melton knew that having a diverse and deep bank of stem cell lines might be critical to what he wanted to do—which was to urge these pluripotent cells to divide and start down the path of becoming pancreatic beta cells. Like children who seem to have a natural propensity for, say, sports or music, some mouse embryonic stem cells were easier to push down this path than others. If the same proved true of human cells, then Melton knew that having enough cell lines available to work with while scientists figured out how to direct this development would be critical. But the government was clearly interested in limiting, not expanding, the number of stem cell lines it would fund.

The are-there-or-aren't-there tug-of-war over the number of lines was a distraction from what he saw as the real issue—that there was no place where anyone could get access to reliable, good quality, robust human embryonic stem cell lines. The government was obviously the wrong place to look; they were still trying to figure out who had what stem cells. Looking for a detour to the roadblock, Melton again took matters into his own hands.

He had to. He now faced another very personal reminder of how high the stakes were for ensuring that stem cell research continued in as rich and rapid a way as possible. Just one month after Bush's announcement, the Meltons learned that their daughter, Emma, fourteen, was also a type 1 diabetic.

It was time, Melton decided, to make his own stem cells.

★ ★ ★

Given the remarkable decisions Melton ended up making to champion stem cell research, you might expect to find something in his upbringing that primed him for such vanguard behavior. But he is the first to admit that there were no such hints of the scientific rebel in his past.

Blue Island, Illinois, where Melton grew up, just south of Chicago, is best known today as a nexus of the city's Metra trains, where six stations untangle a knot of regional lines that converge and split off again to connect the city to its outlying suburbs. While not an island any longer, the city was indeed surrounded by the waters of Lake Chicago when the earliest settlers happened upon the spit of land that, when viewed from a distance, appeared to emerge from a haze of blue mist.

Melton, like so many students of the 1960s, has strong memories of the racial tensions that permeated his Blue Island neighborhood in the aftermath of Martin Luther King's assassination. It's hardly the atmosphere you would picture the soft-spoken Melton in today, but he recalls being a teenager and passing through a metal detector daily at his high school in order to get to class. Tensions reached such a pitch, he says, that his parents decided to move closer to Chicago so that his younger sister and brother could attend a presumably safer school.

"I wouldn't say I got a bad education," he tells me during one of our conversations in his office. "But let's put it this way—compared to what Harvard undergraduates have, it was a different world."

Melton entered the University of Illinois at Urbana-Champaign with the idea of becoming a doctor. Instead, curious about experimental science, he became fascinated with a biology lecture in which the professor introduced him to the concept of limb regeneration—specifically, the remarkable ability of the salamander to regrow a limb that has been cut off.

"Like a little boy, the more I thought about it, the more fascinated I became," says Melton. "I wanted to learn how the process worked."

As an undergraduate biology major, Melton spent a summer in the lab of David Stokum, who was working on limb regeneration. It was there that Melton experienced his first taste of scientific success: He contributed to a scientific paper, and still remembers every detail of the experiment. "It's typical of the things I like to do now, meaning that it's very simple," he says, launching into a synopsis of his study, which involved testing the different ways that a salamander regenerated its limbs, depending on where the appendage had been cut off.

The results, he says, were published "in some second- or third-tier journal. But I was really pleased as an undergraduate to have a study published. That got the science bug in me."

It wasn't until his senior year, though, that Melton truly got inspired. While reading *Scientific American*, he came across a cloning experiment with frogs by a young British scientist, John Gurdon. "Compared to regeneration and what I was thinking about, that experiment just knocked me off my feet," says Melton. "I still to this day can get excited telling people about it. When you tell people about cloning, and explain what it really is, it is surprising to everyone that every cell in the body has all the information to make a whole body. The world didn't have to work that way. That is an unbelievably interesting result, experiment, and fact about life."

Melton promptly looked up where Gurdon worked, and learned that the scientist was based at Cambridge University in England. By this time, Melton had also come to another realization—that practicing medicine wasn't for him. It was the philosophy of science, he thought, that might make a more worthwhile career.

In 1976, he earned a Marshall Scholarship to study at the university of his choice in the United Kingdom. Not surprisingly, Melton chose Cambridge, where he knew Gurdon was continuing his studies on cloning and reprogramming, and decided to complete a second undergraduate degree in history and philosophy of science. But showing the first signs of what would become a hallmark restlessness, Melton quickly came to the conclusion that, contrary to what he had initially thought, becoming a philosopher of science would not engage him enough as a career choice.

"I had a wonderful two-year degree program there; I realized that the most I could hope for, as much as I liked it, would be to write and comment on what other very smart people thought about. I imagine it's like a young boy who wants to be a basketball player," he says, slipping into one of his easy analogies. "At some point, he realizes that the most he can hope for is to go to games, watch the games, and comment on them. But that he will not replace Kevin Garnett."

Melton hadn't forgotten Gurdon, or the frog experiment that had made such an impression on him back in Illinois. So soon after his arrival on campus he made an appointment with the professor and presented the surprised Gurdon with an unusual proposal. Unlike in U.S. universities, where professors head up labs with both undergraduate and graduate students, under the British system, scientists belonged to research institutes and did not typically invite undergraduates to join their teams. Yet here was Melton, asking Gurdon if he could "just help out" in the lab.

"It was the most extraordinary thing," says Gurdon. "He just knocked on my door in the lab, and said he was attending a course in history and philosophy of science and could he spend the weekends washing up [lab glassware] for us. Of course he had a spectacular record, but he didn't choose to mention that at all. He started asking about things, and as time went on, he said he was rather interested in becoming a graduate student. I thought, 'This is no ordinary glass washer.'"

With Gurdon's guidance, Melton absorbed some of the lessons he still applies today in the way he approaches scientific challenges: asking simple, straightforward questions and devising equally direct solutions. Melton left Cambridge with a Ph.D. in molecular biology, having transformed himself from an observer of science into an active contributor.

★ ★ ★

Soon after President Bush's August 2001 executive order, Melton made the trip to Chevy Chase for his annual visit to the Howard Hughes Medical Institute (HHMI). Ensconced in the wealthy suburb of Washington, D.C., the institute is the living legacy of the eccentric billionaire, whose wide-ranging interests led him to create a biomedical research facility that continues to fund three hundred top researchers, like Melton, in an impressive range of basic sciences. Other than producing results, the only condition of the open-ended grants is that the researchers return to the institute once a year to update board members on their progress.

Melton stood in the cavernous auditorium of HHMI's luxurious, wood-enshrouded main building and detailed his work. He described the baby steps his team had made in coaxing embryonic stem cells from mice along the path to creating pancreatic beta cells—the kind that could sense glucose levels and churn out the appropriate amounts of insulin accordingly. He had been able to make the pancreatic beta cells, he said, but saw a journey of considerable challenges ahead in convincing these cells to spurt out insulin, much less be sensitive enough to release the hormone in response to blood sugar.

Melton was following the tried-and-true path of basic science: to figure out as much as he could in the mouse—in his case, teasing out all the pathways and mechanisms involved in making insulin—before moving on to human cells. But even then, he knew there would be years of testing human cells in a petri dish before he could begin applying what he had learned to human patients.

In the months leading up to that day, Melton was increasingly frustrated with the status quo, and impatient with the slow progress that he and others were making. Every day, as Sam and Emma grew older, the potential toll of diabetes would be greater, he knew, as their young kidneys, blood vessels, and other tissues labored under the constant roller-coaster fluctuations of their blood sugar. Sam and Emma couldn't simply join classmates on a whim to kick around a soccer ball at practice, go on a run, or go to sleepovers. They would need to be prepared, anticipating how the new activity would change their blood sugar levels, fueling up ahead of time with fruit juice or a candy bar, and making sure they checked their levels more frequently to ensure they weren't over-anticipating their insulin needs. Watching his children every day, Melton was convinced that the well-trodden path of science would simply take too long to find a cure for diabetes.

So as he addressed the gathered scientists and Hughes officials that afternoon, he proposed taking a shortcut. He was perfectly aware of the many miles he still had to go before understanding beta cells in mice, but he was too tempted by the gift that James Thomson and John Gearhart had bestowed upon the scientific community two years earlier to wait. What Melton wanted to do was detour straight to the human cells in a dish. Before Thomson's and Gearhart's discoveries, he argued, working with the mouse cells had been a necessity, because there simply wasn't a way to isolate and grow human pancreatic beta cells. But with embryonic stem cells, that was now possible. "I used to hold the view that we should wait until we cured diabetes in the mouse, but I guess I changed my view because I'm getting older and life is short," he said.

What he wanted, he explained to Thomas Cech, then president of HHMI, in his office after the presentation, was to make his own human embryonic stem cells, just as Jamie Thomson had.

"He said, 'If I wrote you a letter asking for funding to derive new stem cell lines in collaboration with Boston IVF, would you consider this?'" Cech recalls. "He didn't say, 'Would you promise that you would fund me a lab?' He said, 'Would you consider this?' I said I'd certainly consider it."

Just as Thomson had had a key collaboration with the University of Wisconsin's and the Rambam Medical Center's IVF clinics, Melton knew that he, too, would need access to a supply of good-quality but ethically obtained human embryos. And he had just the partner. Several years earlier, Melton had attended a friend's barbecue and met R. Douglas Powers, the chief scientific officer of Boston IVF.

Boston IVF is a popular chain of IVF clinics in Massachusetts and the leading provider of assisted reproductive services in the state. As the two men talked, Melton described his work using mouse stem cells to generate insulin-producing beta cells, and they eventually fell into a discussion about how to create more such stem cell lines from human tissue. Thomson had used excess IVF embryos to create his lines, and Powers revealed that his clinics alone had thousands of such embryos frozen in storage, which most couples had decided they no longer needed. Never one to miss an opportunity, Melton saw a way to generate the stem cell lines that he was having such difficulty obtaining elsewhere. Working with Boston IVF, he thought, could provide him with a rich supply of embryos from which he could generate these lines. Powers was happy to oblige, and agreed to a collaboration.

With a partner in hand, Melton wrote his proposal to HHMI. A longtime fan of Melton's commitment to science and his sheer determination, Cech meant what he had said. Howard Hughes, as a nonprofit, private institution, was in a unique position to do something about the stem cell conundrum—something that Melton appreciated, if the Howard Hughes board did not, at least not immediately.

With an eighteen-billion-dollar endowment at the time, HHMI was second only to the Bill and Melinda Gates Foundation among nonprofit research organizations, outpacing the Kellogg, Mellon, and Ford foundations in its contributions to research. Melton's request, says Cech, forced the institute to confront the issue that had been swirling in the political arena for months, but that HHMI was consciously keeping at arm's length. When I ask Cech why the institute didn't step in with its considerable resources as the white knight to sustain embryonic stem cell research in lieu of the NIH, he pauses a bit. Stem cell science, I point out, seems to be just the type of cutting-edge research into the "genesis of life" that was the impetus for the founding of the medical institute to begin with. In responding, Cech carefully acknowledges how much the billionaire's initial mission has changed in the half century that HHMI has been operating. "HHMI is not a policy center," he explains. "We are not political. Because there is some power in Washington in being an honest spokesperson for something and not politicizing it." Melton's request, however, brought the stem cell debate right into the quiet institute's serene boardroom and dropped it onto the middle of the table like a ticking bomb.

"We realized of course immediately that this was something we were allowed to do, and we also thought the scientific promise was outstanding," says Cech. "But at the same time, we realized that human embryonic stem cell research, and especially the derivation of new lines, was something that society had a wide range of opinions about. And that this would be a sensitive issue for a number of people, including perhaps some of our trustees.

"Knowing that many of them were Republicans, and strong supporters of the president," Cech tells me, "I wondered what they would say." New to the job of director, he decided to tread carefully. To fully explore the ethical and scientific questions raised by stem cell research, he held a workshop and invited Laurie Zoloth, a respected bioethics professor at Northwestern University who had testified before Clinton's National Bioethics Advisory Commission on stem cell issues, to chair a bioethics board.

Armed with the ethical experts' advice, which included recommendations that donors of embryos be fully informed about how their embryos would be used for stem cell research, Cech was still anxious about how the board would react to Melton's request. But to his surprise, the trustees were overwhelmingly supportive. In fact, after Cech made his case supporting Melton's proposal, Jeremy Knowles, dean of arts and sciences at Harvard, addressed the group to note that not only did the institute have the ability to fund the research, but it had the obligation to do so. The other trustees agreed.

Agreeing to support the research was a far cry from becoming a pro–stem cell, antiadministration lightning rod, however. Such an advocacy-oriented philosophy could not be further from HHMI's primary mission, which focuses more on educating by informing rather than instigating. There is no Howard Hughes Medical Institute facility, for example, other than its administrative headquarters in Maryland. The institute is instead a network of laboratories led by its selected investigators, who are required to hold faculty positions at a university. And in keeping with that position, Cech says, the trustees decided not to highlight their support of stem cells. But at the same time, given its stature and the respect that the institute commanded not just in the scientific field but also in Washington and as a credible and nonpartisan resource for scientific information, "the question came up, should we be open about this, or should we do it privately," he tells me. "We decided we did not want to become the world's spokesperson for embryonic stem cell research. To push this research generally was not our role. But we were proud of our decision, having gone through a rigorous analysis, and we didn't want to hide it."

Supporting Melton's desire to make embryonic stem cells turned out to be a compromise decision for the trustees. What Cech doesn't tell me, but Melton reminds me of later, is that Melton came to HHMI with two proposals. One was to fund his collaboration with Boston IVF to create new stem cell lines. The other, in true Melton fashion, was a broader, visionary appeal to expand HHMI's role in the increasingly noxious stem cell field. With the government, in essence, stepping away from embryo studies, someone had to fill the gap. In Melton's view, that should have been an entity like HHMI.

"I proposed to Howard Hughes that it should take seriously its ability to step in for the government and call for grants to establish a stem cell facility and distribute these cells. I put in a proposal to them which was much bigger and which they declined to do," Melton says. "I am grateful that they supported my research, but they declined the broader approach to the problem."

Melton didn't make his proposal lightly. In fact, he had presented it after considerable thought about a critical question that occupied him as the father of two children who might someday benefit from stem cell–based treatments: Who is in charge of curing disease?

Indulging his education in the history of science, as Melton saw it, five institutions play some role in our attempts to understand and manage diseases, but in the end, only one has a dominant mission of actually *curing* them. Academic institutions like universities, he reasoned, put a premium on learning and exploring, and by definition are primarily focused on laying a foundation of basic knowledge rather than applying this knowledge to some purpose such as treating disease. The same is generally true for government research centers such as the NIH— while many scientists at the institutes are certainly engaged in translating their work to curing patients, by and large, the institutes' mission is focused more on seeking knowledge. What about hospitals? While they are major centers for treating disease, these care centers are preoccupied with treatment, not necessarily cures. Major hospitals now also support research into the diseases they treat, but in reality, their primary business is managing urgent symptoms rather than investing in the longer-term search for cures. The same could certainly be said of pharmaceutical companies—while they are another partner for patients, they are also businesses, and profit incentives dictate that it often makes more financial sense to help patients to manage disease, often over an entire lifetime, rather than cure it.

Which brought Melton to the private foundations and patient advocacy groups. Originally the bit players in the medical research story, in the stem cell narrative, these organizations have gradually shouldered more of both the cost and the responsibility for pushing for cures. "Why did these entities come up?" asks Melton. "My own peculiar view of the world is that it's because the young boy or girl with a disease didn't really have anyone saying, 'Cure my disease.'" Particularly after the Bush policy was put in place, patient advocacy played a pivotal role in keeping stem cell science alive in the United States. It would take many forms—state programs, philanthropic foundations, celebrity organizations—but in the end, all had a similar goal: to promote stem cell research so it could lead to cures.

Which is why Melton felt emboldened to bring his proposal to HHMI. He saw an opportunity for a private institute with HHMI's reach and deep pockets to step in and take the lead.

It may have been too much too soon, however, since the advocacy movement had yet to gain momentum. And given the Hughes's preference to remain outside of the political fray, the trustees, says Cech, were simply not ready to take such a prominent role in such a hotly contested and publicized field. "We weren't going to put out a neon sign that says, 'Stem cells are us,'" he tells me.

But they did agree to provide Melton the seed money for a basement stem cell lab, and to pay a handful of eager young researchers to start making embryonic stem cells.

★ ★ ★

The room measures only 250 square feet or so, barely filling up a quarter of the footage of a standard laboratory. To get there from Melton's office on the Harvard campus, you need to exit the Sherman Fairchild Biochemistry Building, where both his lab and office are located on the fourth floor, and walk a few yards to the red-brick biological research labs. You then take stairs to a basement corridor so rarely accessed that it retains old-fashioned wooden doors etched with numbers in a font decades-old. Across the hall from the machine room, which houses the incessantly groaning generators that power, heat, and cool the rooms above, is your destination. The room that houses the lab itself was previously used for shipping and receiving, and once even served as a machine shop, where operations staff would make minor facilities repairs.

Transformed into a state-of-the-art stem cell lab, the tiny space is divided even further into three rooms. The anteroom is no bigger than a closet; as you swing open the door, it bangs up against a built-in shelf that serves as a desk. The largest room is dominated by a lab bench island occupying the bulk of the central space; it's piled with the paraphernalia of culturing and growing cells—a microscope, a hood, a refrigerator for chilling culture media. Against one wall, running along a small strip flush with the ceiling, is the room's only window, which filters in what little light it can. Against the door are stacked four incubators that house the precious cells that the government refuses to support and that Melton has been generating, de novo, for six years. Those who spent hours upon hours here humorously referred to it as the "escape module." If Harvard ever felt the work being done there was too controversial, they joked, someone would simply just push an Eject button and jettison the entire unit.

The joke reflects a sad truth—the reality of how ostracized stem cell scientists were from the larger scientific community. "What I try to do is create a cocoon," Melton told me during my first visit to the lab. "I tell my students that your job here is to focus on the science. What limits you is your energy and ideas, not what is going on in Washington." By virtue of Bush's decision, virtual walls were being erected that separated the haves—like Melton, who was fortunate enough to benefit from generous sponsors—from the have-nots. For as competitive as scientists are, the practice of science is at its heart collaborative: Researchers build on the findings of others, and the field progresses only when a critical mass of minds and momentum moves it forward.

Bush's order meant, however, that while Melton was theoretically free to collaborate with anyone he wanted in pancreatic cell development, doing so was practically an impossibility. His potential partners would have to be able to separate the time and equipment they devoted to working with his privately derived cells from their government-supported work, which proved too burdensome for any but the most amply endowed labs. Melton's lab became, in essence, a scientific asylum, a stem cell sanctuary that survived and thrived on the self-sustaining energy of the students, investigators, and sponsors that ran it.

On a practical level, the president's order also meant that Melton had to duplicate the equipment for the lab. Because he could not use any federal money to purchase equipment or even pay for personnel who worked on deriving the new stem cell lines, he ended up using his privately raised funds to stock the new space. And just to be sure that there was no cross-pollination, Melton devised green and red stickers for each machine and box of supplies. The green stickers, labeled 'Approved for all hESC,' applied to anything purchased with government grant money and eligible for federally funded studies—these included the presidential human lines, as well as any investigations on mouse stem cells. The red stickers were reserved for anything involving the privately generated lines (lab members naturally became more creative with some of these labels, adorning them with admonitions such as 'Bush says don't use this machine!').

The precautions were an administrative and psychological headache the likes of which Melton and other stem cell researchers had never been burdened with before. They may not have been required by the NIH, but the clean separation of the two types of studies made university higher-ups more comfortable. Jay Lefkowitz, in fact, admits to me that he was especially sensitive to this concern and took it upon himself to alter funding regulations to ensure that work on human embryonic stem cells could go on, legally, in the same labs as federally funded studies (and not run afoul of Dickey-Wicker)—a fact corroborated by other scientists.

As frustrating as the their ostracized status was, however, Melton and his small group of students ultimately found it galvanizing. "The truth is, my lab was never down in the dumps," he says. "I might have been, but they never walked around with their chins on the floor, and never felt that the whole world was against us. We've always had a band-of-brothers mentality."

The lab was Melton's only logical response to the impediments that he confronted from the government and from WARF. With $2.4 million a year from the Howard Hughes Medical Institute and other private philanthropists, Melton kept the lab operating; with a smaller portion of his annual budget coming from government grants, he was able to continue running parallel studies of both federally approved and unapproved stem cell lines from both mice and humans to build a critical base of knowledge about how to make pancreatic beta cells and nurture the first generation of human stem cell scientists.

★ ★ ★

Tall and lean, with a young boy's barely contained energy, Chad Cowan could have been conjured up by central casting as Melton's next recruit. Just completing his doctorate in biology from the University of Texas Southwestern Medical Center as Thomson and Gearhart published their groundbreaking papers, Cowan was enraptured by the promise of stem cells, and immediately saw their potential in moving researchers beyond simple genetic modification to actually understanding human development and disease in an entirely new way.

With Ph.D. in hand in 2000, his first stop on his tour of labs for a postdoctoral fellowship was Thomson's, in Madison. With stem cell science as such a new field, it was the only place he knew of that was doing work on the embryonic cells. It was there that he learned of the handful of groups around the country who were dipping tentatively into the field. Melton's name, however, didn't come up, but others at New York's Rockefeller University did, so Cowan made the trip to Manhattan. It wasn't until he was in the middle of his inquiries there that he finally heard about Melton's new initiative, courtesy of HHMI, in deriving new embryonic stem cell lines.

"My first meeting with Doug was sort of like that famous scene in *Love Story* in the law school library courtyard," Cowan says, smiling, referring to the 1970s film in which Ryan O'Neal first sees Ali MacGraw and the attraction is immediate and electric. "We met, the music played, and all the rest. It seemed like a match made in heaven."

What so attracted Cowan was Melton's vision of the potential of stem cells, and his single-minded focus in ensuring that his lab would be one of the leading sites in the country for pushing the field forward—as long as the end result was the creation of an insulin-producing beta cell. "When I spoke to him, he was like, 'We're going to have thirty to forty stem cell lines and you're going to be able to do all this interesting research with them.' I thought, 'That's going to be fantastic!' I was extremely excited."

About a year later, armed with a green light from Melton and a youthful impatience to launch the experiments he was itching to do, Cowan arrived in Boston. But the honeymoon was already over. "I come to Doug's lab, and lo and behold, we didn't have any stem cell lines," he says.

Cowan's experience was typical of the growing predicament that eager stem cell researchers were facing. While on the face of it Bush's decision meant that the government was funding research on a limited number of embryonic stem cell lines, in reality the cells were proving devilishly difficult to obtain. For academic researchers, this was a startling change from the way they usually traded scientific materials for their experiments, from antibodies to pieces of DNA and even cell lines made from tumors or other tissues of interest. They had always worked on the principle of "If you ask for it, it will come."

But with the stem cell lines, so few people had access to them that there was a mystique building around culturing them just so, and it was slowing the field down. When scientists requested the cells from WiCell, the entity that WARF had established to handle distributing Thomson's cell lines in those early months, for example, they often received excuses rather than cells. "There was an enormous amount of mystery surrounding all of these lines that had all of these different growth conditions," Cowan recalls. "If you asked for the cells, they would be like, 'Well, you know, we'll get that to you as soon as we can; we're having some troubles, it's really hard to grow them.'"

After months of getting stonewalled, Melton finally stumbled upon another way to get his hands on cell lines. Through a colleague in the Department of Genetics, he heard of a young investigator from Israel who was completing a fellowship at Harvard Medical School and had

just begun working with human embryonic stem cells. Nissim Benvenisty, a gentle, easygoing M.D.-Ph.D. from the Hebrew University of
Jerusalem, had trained with Joseph Itskovitz-Eldor, the gynecologist at
Rambam Medical Center, in Haifa, who had collaborated with Thomson. Benvenisty had just begun to use stem cells to study human development and, like Melton, was hoping to exploit them to expose the root
causes of disease. (The Technion stem cell lines were among those tallied
by the NIH and eligible for federal funding.) But because of the long
approval process for launching ES cell studies in Israel, he was temporarily stymied as he waited for his next set of studies to pass muster. In the
interim, he was spending time in the genetics lab of Philip Leder, working on mouse embryonic stem cells. Frustrated, he explained his situation and his preempted human work to a colleague, asking his friend
to keep their discussion to himself, since his findings could not be published until he could proceed further with his work on the human cells.

"Luckily," Benvenisty tells me during a visit to New York, "he did
not keep it to himself." The friend had been a student of Melton's, and
knowing how eager Melton was to work with human embryonic stem
cells, he spilled what he knew of Benvenisty's work to his former mentor. "Doug called me the same day and introduced himself, although
he needed no introduction since everybody knew who Doug was," says
Benvenisty. "And there was instant chemistry. He asked me to come over
to his lab."

That was the winter of 1999, and by 2000, Benvenisty had asked his
lab members in Israel to ship a few frozen vials of stem cells to his new
bench in Melton's laboratory. "Doug got permission from Harvard, and
we started work on the cells the same week," says Benvenisty.

Even these lines, however, which were part of the original batch
that Thomson and his Israeli partners had generated, were not what
they promised to be. A "good" stem cell line, as people were beginning to learn, was one that was relatively young. In stem cell parlance,
that meant one that had been passaged, or split into different plates, no
more than a handful of times. The cells Cowan received were already

worn out; because Thomson had kept them going for a year to ensure their stability, the cells had already been passaged around thirty times. Plus, Thomson admitted to me, he never intended that those initial cells would serve as the source of study material for other researchers; they were merely a proof of concept. "I actually was very conscious of not keeping track of some stuff because I didn't want the hassles," he says. "I didn't keep particular lot numbers and things people might want if they were treating these cells as a resource. But they suddenly became the dominant cell lines because of this funny [political] decision."

The more a human embryonic stem cell (hES) line was passaged, the more likely it was to accumulate signs of wear and tear—alterations in the chromosomes of its cells, any of which could affect the cells' pluripotency and normal division cycles. With normal cell division, cells embark on an exquisitely choreographed program of splitting and fusing its DNA, which is packaged into pairs of crisscross-shaped chromosomes. Before a cell divides, one member of each pair is pulled to opposite ends of a cell as it begins to pinch apart, like a balloon being squeezed at its center, into two cells. The more such divisions occur, the more likely that mistakes in this intricate ballet can arise—especially since the stem cells are being unnaturally arrested, like a needle stuck on a record, in a pluripotent state. Some cells add on extra chromosomes, while others divide without the proper number. The end result is cells that are no longer genetically normal and therefore poor starting material for any further experiments.

When Cowan contacted the few colleagues he knew in the fledgling field who had also managed to work with the approved cells, the experience was the same. Everyone was frustrated. Nobody was happy with the cells, and moreover, at five thousand dollars a vial, they were proving to be an expensive disappointment.

That drove Melton to generate his own bank of lines, starting with the embryos from his collaboration with Boston IVF. In the process, he decided to save his colleagues from suffering the same loss of time and research dollars that he had. He decided to distribute his cells free to

anyone willing to learn how to use them and work with them. As altruistic as it was, he admits to a selfish motive as well. The key for Melton
was to seed the field with enough easily accessible and high-quality raw
material, in the form of cells, to entice bright minds to study them. "I
never believed and still don't believe that a lab like mine can cure a disease like diabetes alone," he says. "I get enormous, essential, absolutely
necessary benefit from many others in all sorts of ways. I don't know
enough to solve this problem by myself."

Cowan was on board as well. "People will now at least be able to
say, 'Here are some cell lines I can use, and here's how to use them.'
We wanted to make sure they were easy to use, easy to thaw, and that
you could get them just by contacting us and saying, 'Hey, I'd like that
reagent,' and boom—it's going to come to you."

★ ★ ★

It was an admirable idea; executing it was a different matter. Although
Cowan didn't arrive in Melton's lab to the incubators full of stem cell
lines that he was expecting, he did have the advantage of collaborating with the husband-wife team of Andrew and Jill McMahon, both
molecular biologists who had considerable experience creating mouse
embryonic stem cell lines.

The key was keeping the cells alive once they hit the culture plate.
As Thomson and Gearhart had learned, keeping human embryonic stem
cells happy in their petri dishes was a 24-7, 365-days-a-year job. Left alone,
the cells would eventually crowd themselves into starvation, soaking up
the nutrient broth of growth factors and enzymes until nothing remained.
But even more important, the cells were itching to differentiate—
that is, to take off on a developmental pathway in which they were no
longer pluripotent and self-renewing, but starting to mature into a specific type of cell, typically a nerve cell, or a heart muscle or skin cell. (It
is one of the ironies of stem cell science that neurons, considered to be
one of the more complicated cells in the human body, with their long,
fingerlike axons that can extend for a foot or more to communicate with

other nerve cells and tissues, are the most easily generated from embryonic stem cells—much to the chagrin of neurobiologists.) For cells, the pluripotent state is an unnatural one, one that is sustained for only a tiny fraction of the human developmental life span. Keeping stem cells stuck in this artificially self-renewing state is akin to revving an engine on an idling car—everything in the system wants to lurch forward; in the stem cells' case, into a more specialized, mature cell.

And for Melton, the stakes were even higher. What he was trying to do was not only culture stem cell lines and keep them in their undifferentiated state, but to do so in a way that was reliable and consistent enough to allow him to pump out large quantities of high-quality cells that could be distributed to cell-hungry researchers around the world. His goal wasn't simply to derive stem cells, but to *manufacture* them.

As Thomson had done, Melton and the McMahons began with a thin blanket of mouse embryonic skin cells known as fibroblasts. Somehow these cells nurtured the human cells, much like a two-dimensional womb. More important, they also managed to keep the cells arrested in their self-renewing state. As Cowan soon discovered, however, the process that Thomson described in his paper for making his successful lines was far from a recipe that just anyone could put on a lab coat and follow. That wasn't unusual in the science world.

"You typically find what are the scarce outlines of how they performed the experiment, and all the baker's tricks aren't included—all the things you might want to pay careful and close attention to," Cowan says. "Deriving human embryonic stem cell lines can take months, so each one of the little things you do along the way could be critical, and you have no idea which one of them is or isn't important. At the time, there was no manual or cookbook for how you do it. Just a paper that said it could be done, and a brief outline."

By the time Cowan arrived, the McMahons had completed the heavy lifting—or, more accurately, the delicate coaxing. After experimenting with dozens of growth cocktails, they had finally homed in on one that seemed to suit the stem cells. Cowan's job was to turn that system into an assembly line. "I came in like Henry Ford, taking something that artists were doing and making it more systematic."

No small task, since growing human embryonic stem cells was proving far more fickle a process than plating dish after dish of mouse stem cells. The culture conditions for growing the mouse cells had not changed much since the days of Martin and Evans, and nearly two decades of practice had made the process relatively routine and almost rote. Researchers had only had three years, by contrast, to hone their skills with the human cells. The process was far more art than science; Cowan faced a formidable challenge in trying to standardize what was at some points akin to "witchcraft and superstition." To make sure that the cells weren't outgrowing their dish, they had to be checked every day. Most stem cell scientists overseeing a new stem cell line will forgo vacations until they can get their line up and running to their satisfaction. "You live and die with these things," confirms Cowan. "When there's a bad day in tissue culture, it's like the kids got into a fight and your entire week is soured as a result."

Like a chef perfecting a new recipe, Cowan spent hours, often alone, in the cramped basement lab mixing and matching a variety of compounds to find just the right culture medium. He threw in expensive growth factors and other elements that he now says proved only marginally helpful in expanding the growth of the cells, but which he decided to include anyway, out of fear that omitting them would doom the line to death. To make the cells heartier, he also had to acclimate them to antibiotics so the cells could be doused with the bacteria-fighting drugs and survive potential contamination.

Once that was done, he spent another three months working out the right freezing and thawing protocol to ensure that once the frozen cells left Melton's lab, they could be warmed up and grown successfully in any lab. "Can you believe it?" Cowan asks me, laughing now over what he considers to be the trivial nature of the experiments. "They gave me a Ph.D., I'm at Harvard, and I'm working on freezing and thawing stem cells."

But as routine as it seems now, it was a monumental, unanswered question back in 2004. Nobody knew how best to freeze the delicate lines and then bring them out of their frozen slumber. What culture medium would flash-freeze the cells and keep them metabolically intact upon thawing? Was it better to thaw in stages or, like removing a bandage, all at once, with a blast of heat?

The result, after nearly two years and hundreds of hours in the escape capsule, was seventeen human embryonic stem cell lines that met Melton's strict criteria for being robust enough to distribute to labs around the world. In March 2004, Melton's team published a paper in the *New England Journal of Medicine* outlining how they derived their lines, along with an open invitation to scientists to come and get 'em through a standard Material Transfer Agreement that contained none of the more onerous strings that trailed along with the Wisconsin lines. "I don't ask people what they are going to do with reagents I send them," says Melton. "I figure it's none of my business. If they beat me to an important discovery, well, more power to them."

In a commentary in the same issue of the journal, John Gearhart, who along with Thomson had generated the first human versions of the lines, lauded them as an "important contribution" and a "tour de force."

In the first month, dozens of requests flooded into the tiny anteroom of that basement lab, laying the foundation for a cottage industry for stem cells, in which revenues were measured not in dollars but in knowledge. Along with the cells themselves, Melton's lab sent along a care package of everything researchers would need to get started on stem cell experiments. Included was Cowan's stem cell bible, a detailed set of guidelines that outlined how he and his team had grown up the culture, under which conditions, and how to replicate those conditions to keep the cells happy. It was the cookbook that Cowan had lacked when he began, and like any good manual for the kitchen, it included the helpful hints that only an experienced hand could pass along—tips such as warming all the dishes up to the cozy 37°C temperature found inside the incubators before placing cells on them, or sliding plates from side to side instead of swirling their contents to distribute newly transferred cells, and using prechilled vials when prepping cells for freezing. Because the Bush policy had limited the number of people able to derive new stem cell lines, Harvard, or more specifically, the Harvard Stem Cell Institute, the entity that Melton cofounded to virtually gather the university's stem cell efforts into a single program, became a mecca for anyone interested in learning how to make and maintain hES cells.

"There were no state secrets," says Cowan about the Melton lab's stem cell efforts. "It wasn't like, 'Oh, we're going to get a leg up on the whole field because we've got this extra expertise.' There was none of that. Doug never thought like that. It was more of an altruistic, 'Everyone should have access to this because it's all-empowering and because it's one of these technologies that changes the game.'"

President Bush had written the rules; Melton had rewritten them. Now it was just a question of who would come out to play.

7

California Dreamin'

Like many Americans in the summer of 2001, Robert Klein was only vaguely paying attention to the buzz over stem cells in Washington. He was still blissfully unaware of how deeply he and his family would soon be affected by the fate of the still nascent field. The hearings on the Hill, the appeals to the White House, the headlines in the national newspapers, the news flashes on broadcasts, as well as the building angst among the scientists were, at the time, all an interesting but still distant concept to Klein. For his son Jordan's sake, Klein's attentions were focused on the more immediate strategies that doctors were testing to help diabetics—experimental approaches such as the Edmonton protocol of transplanting new insulin-producing cells into patients so they could begin making sufficient quantities of the hormone and perhaps even lower the number of injections they needed each day.

"I had no idea," he says of the debate over stem cells at the time. And as for Bush's August 9 decision limiting federal funding of the research, Klein says, "I didn't understand that a) it was a gross misrepresentation [of the state of the field]; and b) how critical stem cell research was going to be in curing disease."

Klein has met me in a café near his Portola Valley home in Palo Alto, California, and despite fighting a cold, he is eager to talk about the subject that has consumed his life since that August day in 2001. Tanned, with a head of slick, wavy, graying hair, he has an easy smile and an executive's confidence. He looks you in the eye, and isn't afraid to take on difficult issues—or the critics who frequently describe him as a bully or an egoist.

There's no denying that Klein is one of those people who leave a strong impression. Before meeting him, I notice a pattern in the way people who knew and worked with him always talk about their colleague. It would go something like this: Yes, they would acknowledge, he had prodigious business skills, and was adept at fund-raising and managing what would become a history-making effort by the state of California to fill the yawning gap in stem cell research that President Bush's executive order had created. They would go on to credit him with turning an intriguing idea—using a state's tax revenues to fund a new field of science through the sale of bonds—into a reality, describing him as brilliant, charismatic, and "an unstoppable force." But more often than not, at some point the compliments would quickly turn into a "but," as in, "but he has an enormous ego," or "but he wants too much control," as they describe his frequently abrasive personality and domineering leadership style.

Growing up in Fresno, where his father served as city manager, biology was an anathema, and advocating for medical research was certainly not on his short list of career interests. But complex problems—the kind that require navigating the intricate and often arcane webs of policy and law, politics, and state and municipal bureaucracies—lured the young Klein in. Armed with a law degree from Stanford, and brimming with youthful confidence about the prevailing power of justice and equity, he and Chris Miller, a longtime colleague, befriended a woman whose life bore little resemblance to his own privileged upbringing. An African American mother living in Orange County, Mary Williams was battling alcoholism and poverty while trying to raise seven children and facing eviction. Klein, inspired by a law school professor who had taught a memorable course on low-income housing, decided to become her benefactor, along with Miller, and they ultimately guided Williams through the red tape of the public housing bureaucracy until they were able to secure an affordable home for her and her children.

That experience dropped Klein squarely into the 1970s California housing market, where demand, particularly for low- to middle-income housing, far outstripped supply. Seeing an opportunity, he and an associate, Michael BeVier, turned government subsidies into a lucrative series of housing projects. To overcome the constraints of the limited subsidies, the pair combined the financing of market-rate properties with the subsidized ones, and generated returns that could be funneled back into supporting the public housing projects. But in 1973, the Nixon administration abruptly ended the subsidies, and Klein confronted a brusque end to his budding real estate career.

Yet in a display of the innovative thinking that he would make a hallmark of California's stem cell program, he turned to bonds. To bypass the federal block of funds, he and BeVier went to the State of California and became consultants for the legislature as they crafted a bill that, for the first time, relied on the sale of bonds to subsidize housing projects. The bill established the California Housing Finance Agency, the state's affordable-housing lender for low-income and first-time home buyers that backs close to fourteen thousand mortgages each year for more than one billion dollars.

That figure certainly set a precedent in the state at the time, but even that sum seems paltry compared to what Klein ultimately convinced state voters to invest in the California Institute for Regenerative Medicine (CIRM), the groundbreaking stem cell research agency that he helped to establish in 2004, using a similar bond-based financing scheme that cost six billion dollars.

★ ★ ★

When President Bush established August 9, 2001, as the cutoff date for federally eligible stem cell lines, stem cell experts—and patient advocates—could do one of two things. They could make do with what ultimately ended up being only about two dozen viable stem cell lines. Or they could find a way to make more.

With NIH bankrolling more than 90 percent of all biomedical research in the United States, however, the second option was essentially a nonstarter. Despite the hefty coffers of nonprofits such as the Howard Hughes Medical Institute and university endowment funds, the sheer volume of money needed to support a growing field of science was simply out of reach for anyone but the government. Even states, although eager to step in, did not have the financial means to sustain a viable research effort for something as complex and controversial as embryonic stem cells. And that was even if residents thought the research was ethically acceptable—some states, such as South Dakota, ban any research on embryos, while others, like Nebraska, limit use of state funds to certain embryonic stem cell studies.

Even so, with the void left by the NIH, states seemed to be the likeliest candidates for picking up the stem cell mantle. The only question was how.

Depending on whom you ask, you'll get a slightly different version of the story about who came up with the idea of bringing stem cell research to California. Klein recalls a phone conversation with film producer Janet Zucker, who with her husband, Jerry (of *Airplane!* and *Ghost* fame), and friends Doug Wick and Lucy Fisher had created a grassroots organization to support stem cell research. The producers came to Klein with an idea for asking California citizens to fund stem cell studies in the state by pushing through a request for one billion dollars in appropriations. "I told them there's no way you're going to get one billion dollars through the state legislature," Klein tells me. "It's just not there; there are too many other critical needs—public clinics, public hospitals going bankrupt; this is not the right way to finance it." The right way, he told them, was to use bonds. It had worked before, he knew, to finance low-income housing, and he couldn't see why it wouldn't work again.

The Zuckers, however, and other prominent scientists recruited in those early days of the movement tell a different story, and place the seed of the idea nearly ninety miles to the northeast, in the offices of a state senator in Sacramento. "A lot of other people take credit for it," says Dr. Larry Goldstein, a stem cell researcher at University of California, Irvine, who became an early supporter of the state-backed plan. "But as far as I'm concerned, I would trace it back to Deborah Ortiz and her staff who really got the thing rolling."

★ ★ ★

Gregarious and articulate, Ortiz was by 2003 a seven-year member of the state legislature, having served both in the state assembly and, since 1998, in the California senate. Medical research was a particular interest of hers at the time—her mother had battled two different cancers, and after a three-year struggle with ovarian cancer, had recently passed away. For nearly four years, Ortiz had spearheaded an effort in the state legislature to appropriate eighty million dollars for prostate and ovarian cancer research, a measure that required a two-thirds vote of the legislature each year.

"People had high hopes for embryonic stem cell research at that time," Ortiz tells me during a phone conversation from her office in Santa Barbara, where she is currently vice president of public affairs for Planned Parenthood. "And it was their sense that this was the only area of research facing a ban. So for that reason, we thought we must carve out something special to it."

In Washington, taking cues from Bush's new policy, Senator Sam Brownback of Kansas was urging Congress to criminalize research on embryonic stem cells—his bill attached a one million-dollar fine and ten years in jail to any scientist engaging in embryonic stem cell research. The penalty applied as well to any researcher importing cells from abroad, and even looped in patients who might fly overseas, receive stem cell therapy, and return to the United States with those cells in their body. All were equally subject to the penalties.

Closer to home for Ortiz, Roger Pedersen, the UCSF biologist who along with James Thomson and John Gearhart had been in the race to make the first human embryonic stem cells, had just left the country for England, citing his worry over being able to continue his research on embryonic stem cells in the face of increasingly tighter restrictions and the threat of jail time.

"There was also the element of really wanting to make sure this was an emerging area that we could have a pipeline of research doing fascinating, early—very early—stage research that might have some promise," says Ortiz. "And if we didn't create some hope of stem cell scientists being able to continue in the field—in California, ideally—then would we cut off the pipeline of potential researchers going into this area of research, of embryonic stem cell research?"

California seemed the ideal place to launch such a grassroots movement. The state was already home to a formidable cadre of scientists, including a sizable share of Nobel laureates, owing to the network of state-supported research universities in the University of California system, as well as private institutions such as Scripps and Stanford University. There was also a dense population of biotechnology interests in the state, many of whom already had strong relationships with academic centers to translate the basic research findings into usable commercial products. These scientists were all raising alarms over the intrusion of political ideology into the practice of good science. By limiting government support and involvement in the field, they argued, the United States was essentially starving it to death.

Ortiz wondered whether there was a role for states to play. A policy wonk, she began toying with a possible exception to the supremacy clause that established federal dominance over states when it came to amendments such as Dickey-Wicker. If the federal government prohibited embryonic cell research, banning it altogether, then states would have to comply. But states are entitled to a narrow leeway when it comes to protecting the health and safety of its residents. Wouldn't stem cell science qualify as just such an exception? Could states essentially start a grassroots movement, throwing down the gauntlet to the administration in an act of defiance?

Ortiz, along with many legislators and scientists, was convinced that the political winds would eventually shift to loosen the presidential restrictions. But if states in the meantime could step in to fill the gap left by the NIH, then maybe, just maybe, the field could be sustained, albeit on life support, until that time. "They liked the idea that if anybody could do it, California could do it," she says of her staff's consensus on the matter.

But before anything could be done with embryonic stem cells, the field itself needed a thicker skin. With assaults from Brownback and the Dickey-Wicker amendment still in place, Ortiz knew it wasn't enough to simply invite researchers to California's sunny shores to do their work. They needed a safe haven, and she gave them that with a bill permitting therapeutic cloning, the research version of what Wilmut had used to create Dolly. Under therapeutic cloning, no human embryos would be used to clone another human being; instead, the embryos would only be allowed to develop to fourteen days, at which point their stem cells would be removed, effectively stopping further development. It made California the first state to approve the research, also known as nuclear transfer, and Governor Gray Davis signed the groundbreaking legislation into law in September 2002, nearly a year after Bush's executive order.

Welcoming research, however, was not the same as making it possible. For that, the state needed dollars—lots of dollars. But unwilling to jeopardize her annual appropriations for ovarian and prostate cancer research, which had bipartisan support in the legislature by adding another tax-based obligation, Ortiz cast about for a way to circumvent the process in Sacramento and go directly to the voters.

Why not, she thought, take the stem cell issue to the streets, pound the pavement and knock on doors to gather signatures and get the citizens of the state to support funding the science? It wasn't an entirely crazy idea, since such initiatives had become a popular if quirky feature of California politics. But when she took it to some of the state's leading scientists, they weren't so sure.

Irv Weissman, the developmental biologist at Stanford University who had isolated the first human adult stem cell, in blood, expressed reservations over placing such a morally controversial issue in front of the public. He wanted to ensure that people understood the difference between therapeutic cloning, which the scientists were proposing as a way to access ES cells, and reproductive cloning, or the making of cloned babies, which most researchers opposed. And he didn't think that a ballot buried in the list of initiatives in a political campaign was the proper place to make this distinction. Weissman was worried that if people believed that the ES cells came from aborted fetuses, then the measure would get mired in the quagmire of abortion politics. Not being as steeped in politics, however, as Ortiz's staff, Weissman was surprised by the response of her political consultant when he raised his concern. "He said you *want* to have it linked to abortion because in California, 60 percent of people believe it is a woman's right to have an abortion," he says, smiling at the logic. It was the first in a long list of lessons for Weissman in the politics of science.

★ ★ ★

It didn't take long for Ortiz's efforts in Sacramento and the Zuckers' group in Los Angeles to converge.

And in typical LA fashion, it turned out that it was all about whom you knew. Weissman's sister, Lauren, a movie studio executive, was serving as the director of CuresNow, the stem cell advocacy group that the Zuckers, Wick, and Fisher had created to lobby for stem cell research. By this time, the Zuckers had logged several trips to Washington, and while they had contributed to defeating the Brownback bill, they realized that they had only succeeded in putting out a fire; the embryonic stem cell issue was more like a smoldering ember, and they needed something far more substantive and consistent to effectively douse the opposition. They began to discuss bringing their efforts closer to home—if it wasn't going to happen on the national level as long as the Bush administration remained in the White House, then perhaps they should concentrate their activities in California.

Thanks to Ortiz's legislation, that strategy seemed reasonable. "Lauren called me one day, and she said [she] was talking to Deborah Ortiz and [Ortiz] said, 'Why not do a bond initiative in California?' " says Jerry Zucker. "I said, 'God, that's incredible, can you do that? Is that real?' "

If Klein tends to rub people the wrong way with his take-charge personality, Jerry and Janet Zucker have the opposite effect. Just as passionate as Klein is about stem cells, the Zuckers are immediately warm and inviting, and opened up their home to the first brainstorming sessions that eventually gave birth to Proposition 71.

What the Zuckers and Klein had in common was a child with type 1 diabetes.

Jerry Zucker will never forget the day that he, his wife, and their eleven-year-old daughter, Katie, confirmed Katie's diagnosis, for the way it tore his emotions from the heights of gratitude to the depths of fear. "She hadn't been well for a while," he tells me during a candid, hours-long conversation. "She had all the usual symptoms—she had been losing weight, she had chapped lips, and no energy. But [when we took her to the hospital] they put a needle into this little vial of liquid, of insulin, and injected it into her and all of a sudden, within twenty minutes, we had our daughter back. As difficult a day as that was, it was also kind of euphoric in a sense, because she was running around, giddy and smiling. . . . But," he adds after a pause, "you realize that if your daughter is ever going to find a cure, it's not going to come from a pharmacy or a doctor; it's going to come from a scientist."

It was that realization, that it would be basic science that would relieve Katie from her lifetime of dependence on glucose monitors and insulin injections, that made the Zuckers so incensed over the Bush administration's restrictions on stem cell research. Like many parents with children affected by the disease, the Zuckers saw the limitations as a door being shut in their daughter's face and those of millions like her who might benefit from science that wasn't being encouraged. Particularly galling for them was the argument that IVF clinic embryos that were going to be destroyed could not be used for research purposes. "That's what drove me crazy when Bush first appeared and said, 'I'm going to allow use of stem cells that are already stem cells and in which the embryos have already been destroyed, but not let any additional embryos be destroyed,'" Jerry Zucker says. "It seemed so Solomon-like to everyone, but it was idiotic."

So when Lauren Weissman called Jerry, he was ready to listen.

Together, Ortiz and the Zuckers held a series of dinners that formed the core of what would become Proposition 71, California's stem cell initiative. In Sacramento, and at the trendy One Market restaurant in San Francisco, and even at the Zuckers' Brentwood home, the state's leading scientists and stem cell advocates gathered over the course of the next year to craft an action plan. They discussed the strategy of selling bonds to fund the research, and dreamed of raising enough to attract leading stem cell scientists from around the world to make the state a beacon and a model for circumventing the federal ban, and that other states might follow. "We knew nothing; we knew zero," Jerry Zucker admits of those early discussions about getting their proposal on the ballot.

So despite the good intentions, for the most part the meetings were high on energy but low on any actual action. "You could see there was a lot of earnestness, a lot of feeling of 'we've got to do something,' a lot of people saying we've got to do it politically, but no single leader who would galvanize all the forces required to get there," says Weissman, who attended several of the gatherings.

Also at some of the early dinners was Paul Berg, Weissman's colleague at Stanford, who was tapped to provide counsel and scientific gravitas to the effort. A biochemist, Berg had not worked directly with stem cells, but he had earned a Nobel Prize for his pioneering experiments in the 1970s in splicing and recombining DNA. Diabetics owe a debt of gratitude to Berg for his innovations, which led to a synthetic form of insulin that eliminated the complications and cost of animal-based insulin.

A thoughtful, deliberate man with a grandfatherly demeanor, Berg is widely viewed as the father of genetic engineering. But it wasn't merely his scientific contributions that made him an ideal recruit into the stem cell ranks—Berg is also respected for his efforts in devising a remarkably effective and proactive response to the social and ethical specter that his own work in the lab created. Because his technique of recombining DNA made it possible to mix and match snippets of DNA, or genes, among species—splicing together human genes, for instance, into a bacteria in order to better understand their function—Berg and his colleagues recognized that the science was ripe for abuse. There was a very real possibility that man-made genetic horrors could harm not just the human species but the environment as well. So in 1975, Berg took the unusual step of convening a group of genetic experts at Asilomar, in California's Pacific Grove, to address the safety concerns the new science presented. After several days of discussion, the brain trust, which included James Watson, the codiscoverer of the structure of DNA, as well as past and present Nobel laureates, decided on a voluntary moratorium on certain types of recombinant studies.

Known as the Asilomar Conference, that gathering remains a highlight of scientific policy-making. It was an example of effective self-policing, and the recommendations of the attendees later formed the basis of guidelines established by the National Institutes of Health on the appropriate use of recombinant genetic techniques in research.

With stem cells, some researchers were beginning to see traces of the same broader worries over the potential for science to run amok—and the same concern that this social pressure would start to suffocate the science. With his experience from Asilomar, Berg, the stem cell advocates felt, was an ideal ally.

Like Weissman, Berg sensed the same well-meaning energy of the group, yet realized they needed a more defined mission. "They tried to get enough of a wave of activity and feeling against the president's policy, by going to their Congresspeople and so forth. But it didn't make much headway," he says.

After attending a few of these meetings, it was Peter Van Etten, of JDRF, who realized what the group needed. It needed Bob Klein.

★ ★ ★

When Van Etten called Klein, he caught him in the middle of preparing for another trip with his daughter—"same daughter," Klein tells me with

a smile, as the one with whom he was traveling when Jordan was diagnosed four years earlier. This time, they were celebrating her graduation
from college with a trip to Peru, to spend some time in the Amazon
jungle.

For Van Etten, Klein's tenacity, as well as his growing connections to
both political and financial leaders in the state, was exactly what the Hollywood group needed to take their effort to the next level. So he asked
Klein, whom he had come to know as an active member of the Palo Alto
chapter of JDRF, to speak to Janet Zucker.

According to Klein, he spent an hour and a half on the phone with
Janet during that first exchange, outlining a bond-based financing
scheme in which he argued that the best way to finance the research
was for the state to sell bonds, but delay payment of the interest on the
sales for a few years. The idea was to amortize the burden over several
generations, since investment in medical research, unlike that in bricks-
and-mortar projects, would not yield any products or benefits for years, if
not decades, to come. That way, the burdens and benefits of the research
would be more equitable among those investing in the science and those
who might ultimately enjoy the products of that investment.

If many of the original backers of Proposition 71 don't agree on who
actually came up with the idea of using bond financing, they do credit
Klein with helping to sell it to the public. While the concept of using
bond financing to fund an entity as amorphous as the promise inherent
in a new science was certainly unprecedented, Klein argued that investing in medical research, so intangible and uncertain, is in a sense no
different from investing in housing projects or highways or schools—as
long as citizens accept that the capital in question is intellectual rather
than concrete. "I said, 'Look, they need to understand that the medical
research that is being done is intellectual capital—it's not operating cost.'
If they understand that it's one of the most important sources of capital
in a society, they'll understand that you can do bonds to finance it. But
people have to make the paradigm shift in understanding that this is a
capital asset of society that protects the society from massive suffering

over decades. It's like amortizing the asset that you would invest in a bridge or harbor that continues to yield returns over decades."

Klein was in his element, and it was clear his experience and energy would be an asset to getting the initiative past voters. "He essentially took it over," says Ortiz, with just a trace of regret. "He clearly had greater wealth, although we could have done it, I think. It essentially became something he then controlled."

★ ★ ★

Klein's first hurdle was deciding how much money it would take to fund an entirely new field of science. Tens of millions? Hundreds of millions? One billion dollars, as Ortiz's staff had proposed? Several billion? At a dinner at Klein's Portola Valley home, he and Van Etten played around with some numbers. After tallying up the costs of previous research projects, and taking into account how much the NIH's $250 million investment in human embryonic stem cell studies had yielded, they came up with three billion dollars to be paid out over ten years. Add on the interest, and the total came to a head-spinning six billion dollars.

Even Weissman, who advised the group and provided scientific perspective on what would need to be done to launch and support a serious, robust program in embryonic stem cell research, admits that the number surprised him. "It made my jaw drop," he says.

Klein envisioned that the money would support not just grants for research studies on all types of stem cells, with a priority on the embryonic stem cell studies that the government would not fund, but also the construction of new stem cell–focused facilities to house the state's growing scientific efforts. The funding would be split into $295 million a year over ten years, and was twelve times the amount that the NIH had spent on embryonic stem cell research in 2003. It also made the figures that presidential candidates George Bush and Senator John Kerry were pledging for the field at the time—twenty-five million and one hundred million, respectively—look ridiculously inadequate.

But nobody knew yet whether the citizens of California would be willing to part with three billion dollars of their taxes to support the field. In April 2004, the nearly bankrupt state was forced to approve fifteen billion dollars in deficit funding bonds in order to remain solvent. Klein himself was beginning to doubt whether the plan would work. "It was not a great year to have an initiative on the ballot where you would need to include six billion dollars in stem cell research," he says.

By late spring, the campaign was eating up one million dollars per week, and in those early days only a handful of supporters were willing to pony up for the cause. Three families composed the core of the Prop. 71 group: the Zuckers, the Kleins, and Tom Coleman and his wife, Polly, friends of Klein's who also had a child with diabetes. The Zuckers and the Colemans financed a number of polls and were bankrolling the effort to educate Californians on what stem cells were and why they needed to support the state's effort to fund the research. At one point, when the coffers dwindled, Klein put his house on the market to keep the campaign alive. "That was a pretty tough place," he says. "I needed to raise a million dollars a week in non-deductible political contributions. We were running out of liquid assets, and other people weren't stepping up."

For Klein those months in the beginning of 2004 were the darkest of the entire campaign. For his partners, it was a turning point as well. As the groups struggled with keeping the campaign going, tensions began building internally. What had, according to Jerry Zucker, begun as a Three Musketeers all-for-one culture was rapidly souring into a power grab on Klein's part. When it came time to form a corporation to run the initiative's campaign, all agreed that Klein would convert his nonprofit organization, California Research and Cures Coalition, into the leading organization for the campaign. But the normally easygoing Jerry exploded when Klein took it upon himself to appoint his ex-wife and his assistant to the board. "I remember having a screaming conversation with Bob," says Jerry, "about some issue that needed to be decided, and Bob simply said no, and because he controlled the corporation, he could do what he wanted." Klein, he says, justified his ex-wife's role because much of his assets were still tied up with hers, and he included his assistant, whom the Zuckers and the Colemans both respected, as a reward for her hard work on their behalf.

There was another showdown at the group's offices, for which political consultant Chuck Winner, who was responsible for the message of the "Yes on 71" campaign, played uneasy referee. "Bob wouldn't back down," says Jerry. It was a turning point for him and his wife, as well as for the Colemans. "Janet and I really had to make a decision, at that moment, on what was more important," he says. "We decided, we can't trust this guy and he's not an honest guy, but he is a forceful guy and we need him to pass this thing. We'll just take a backseat now, and it's his football. We'll play by Bob's rules and just try to yell and scream a bit for what we think is important."

★ ★ ★

Perhaps the only bright light during that time was the emergence of a very important voice in the campaign, just when it was needed most. Ever since her husband had been diagnosed with Alzheimer's disease, Nancy Reagan had come to appreciate the potential of stem cells to expose the mysteries of incurable diseases. Despite her husband's well-known conservatism and antiabortion position, Nancy had been taking ever bolder steps in support of the science that she increasingly believed could help millions of patients, despite its dependence on embryos. In 2001, she had written a plea to President Bush to support embryonic stem cell studies. Just a few years later, she appealed to Senator Orrin Hatch, another Republican with strong views against abortion, but who favored expanding stem cell research. When Hatch proposed legislation in Congress that would ban reproductive cloning but allow therapeutic cloning in order to generate embryonic stem cells, he was no doubt influenced by Nancy's personal request. In describing her husband's slowly deteriorating mind, Nancy wrote, "Because of this, I am determined to do what I can to save other families from this pain. Orrin, there are so many diseases that can be cured, or at least helped, that we can't turn our back on this."

But it wasn't until May 2004, as CuresNow was launching its effort to gather enough signatures to put their stem cell initiative on the ballot, that Nancy took her appeal public. At a fund-raiser for the JDRF at the Beverly Hilton hotel to honor her courageous support of the field, she made her first public statement in support of the science. Walking to the podium in tears to accept her award from Michael J. Fox, who had been diagnosed with Parkinson's in 1991 and is a vocal stem cell supporter, Nancy described the pain of watching her beloved husband slip away, and repeated the plea she had made to Hatch. "Ronnie's long journey has finally taken him to a distant place where I can no longer reach him," she said. "Science has presented us with a hope called stem cell research, which may provide our scientists with many answers that for so long have been beyond our grasp. . . . We have lost so much time already. I just really can't bear to lose any more."

When Reagan died barely a month later, Nancy's words took on an added poignancy. Two days before his death, California's stem cell initiative qualified for the ballot in the November election. For the first time, a state would go to the polls to decide whether or not to fund stem cell research.

★ ★ ★

Ortiz tells me she knew her billion-dollar appropriations bill would not garner the two-thirds majority support to make it through the legislature, but that she wanted to put the bills out there in order to work out the financial, legal, and policy issues. She saw the debate as a way to air out the financial structure that would be required to take the unprecedented step of having a state step in and fund an emerging field of science without holding the federal government's hand.

As it turned out, the groundwork laid by both the legislation Ortiz managed to pass (making California the first state to allow embryonic stem cell research) as well as those that didn't (the billion-dollar appropriation to fund the work) may have given Prop. 71 a leg up with the voters. In August 2004, in the first poll on voters' attitudes about the initiative, Californians were nearly split, with a slight majority, 45 percent, voicing support of a state-funded program in ES cell research and 42 percent opposed.

By October, the Prop. 71 campaign had broken out of its dry spell and had spent $21.5 million canvassing the state with ads, stump speeches, and science seminars. Many noted that the effort to get the proposition passed looked like a political campaign for its organizer, Klein.

But no one could deny his fund-raising abilities. As the initiative gained momentum, a who's who of deep-pocketed backers joined the cause, including Microsoft founder Bill Gates; Gordon Gund, owner of the Cleveland Cavaliers and a financier who had lost his eyesight to retinitis pigmentosa; eBay founder Pierre Omidyar and his wife, Pam; as well as a slew of Silicon Valley venture capitalists who collectively contributed $7 million. The JDRF contributed $1 million. Klein threw in an additional $3 million of his own. The opposition, namely women's rights groups who voiced concerns over the potential exploitation of women for their eggs, and the Catholic Church, mustered a paltry $200,000 by comparison.

And since it wouldn't be a California campaign without celebrities, stars from Brad Pitt to Michael J. Fox filmed television campaign advertisements urging people to vote yes on 71. On the last weekend before the November 2 vote, a posthumous television spot of actor Christopher Reeve, a tireless stem cell advocate after he became paralyzed in a 1995 horseback-riding accident, aired throughout the state. Reeve, who had passed away just a month earlier, had been a supporter of Prop. 71, and his appeal was preceded by a message from his family noting that the statement was his last recorded message. "Please support Prop. 71," Reeve urged in the spot, between breaths on the ventilator that kept him alive, "and stand up for those who can't."

On November 2, 2004, California citizens went to the polls, and by a margin of 19 percent approved a landmark initiative, the first state-funded investment in a scientific field. At a victory celebration at the Beverly Hilton in Los Angeles, Klein took center stage and wasted no time reminding the audience of the historical significance of Prop. 71. "All of us could have lived a hundred lifetimes and not had the opportunity to change the future of human suffering as the people of California have tonight," he told the hundreds gathered in the ballroom.

Oritz, who showed up but was initially denied entry to the party, was eventually brought up on stage by Jerry Zucker, who introduced her as the architect of Proposition 71.

★ ★ ★

No sooner had the initiative passed than the glow of victory began to dim under a harsh cloud of criticism. Opponents of Prop. 71 shrewdly couched their complaints not in ideological terms—as Ortiz's political adviser had pointed out, California is one of the most liberal states in the union when it comes to abortion and reproductive rights—but in fiscal ones. The state, after all, was nearly bankrupt and needed a loan to stay afloat, so the critics took their case to court. Arguing that the agency created by Prop. 71, the independent CIRM, was unconstitutional because it was illegal for a non–state entity to dispense state funds, People's Advocate and the National Tax Limitation Committee called for the disbanding of CIRM. Another suit, filed by Californians for Public Accountability and Ethical Science, charged that CIRM inappropriately exempted its members from state conflict-of-interest laws.

It didn't help matters that the new chairman of the board of CIRM was Klein, by unanimous nomination. According to the intricate selection rules he had written into the proposition, the governor, lieutenant governor, treasurer, and controller were each required to nominate candidates for chairperson and vice chairperson of the Independent Citizens' Oversight Committee, the board of the CIRM. The remaining twenty-five members of the board would then vote on the candidates. To few people's surprise, however, all four nominated Klein for the top post.

Still, in the end, even Klein's detractors admit that he may have been the best man for the job, given the legal and legislative challenges that the fledgling agency faced in its first two years. Before any grants could be requested or any science could be funded, CIRM had to navigate the lawsuits and overcome several attempts by the California state legislature to alter its structure, all the while taking on the Herculean task of essentially re-creating a Golden State version of the NIH.

These were, however, just the kinds of challenges that Klein relished. He had, after all, written the initiative anticipating just such efforts to topple it. Which is why, he tells me with a grin, he wrote Prop. 71 as a constitutional amendment, carving a place for stem cell research, and CIRM, right into the California constitution. Some say the proposition itself is the longest one on the California books—it stretches across eight and a quarter pages, and it's highly unlikely that too many voters in the booth on that election day in November 2004 read their way through the entire text. If they had, they'd have seen at the very end, closing out the proposition, perhaps the most important feature of the new law: To make any changes to the proposition or the agency it created, legislators would have to do so with a "bill introduced and passed no earlier than the third full calendar year following adoption, by 70 percent of the membership of both houses of the Legislature, and signed by the Governor."

As Ortiz and anyone else familiar with California politics knows, that was a virtual guarantee that the initiative would remain intact; it was, as many described it, "bulletproof."

"It was written to insulate it from state politics," Zach Hall, a neurobiologist who became CIRM's first interim president, tells me over breakfast one morning in New York about the carefully crafted proposition. "One of the things that infuriated a lot of people—legislators and others—is that it's almost impossible to change anything. But it also meant that even desirable changes couldn't be made."

Hall, in fact, learned that the hard way when he took the helm of CIRM in 2005. A smart, well-respected scientist with considerable administrative experience, Hall literally walked into the job when he wandered over to the second meeting of the agency's board on the University of Southern California campus in Los Angeles in January 2005. After heading the NIH's National Institute of Neurological

Disorders and Stroke from 1994 to 1997, and following administrative stints at the University of California, San Francisco, and USC, Hall was living in San Francisco but commuting to Pasadena, near Los Angeles, where he was director of the neurogenetics institute at the medical school. Eager to cut down on his commute, Hall was considering one last job in the Bay Area before retirement when Ed Penhoet, vice chair of the new stem cell agency, asked if he would consider the CIRM presidency. Learning that the term was six years, Hall said no, adding, "I don't want to work that long."

But after sitting through the five-hour meeting, it was clear that the agency, which Hall believed in, would need to get started quickly, and that the board wanted to install an interim president while it searched for a permanent one. So he let Penhoet and the board know that he was interested.

As Hall quickly found out, however, there was a considerable blurring of the lines between the duties of the chairman of the board and the president of CIRM, a point of ongoing tension between him and Klein during his two-year tenure. "When I arrived, I found that with one exception, everybody on the staff were Bob's campaign people who had been brought in to run the agency and who still reported to him more or less," says Hall. "It took a while to sort that out."

Hall's was a thankless job. While the litigation against CIRM dragged on, the agency was prevented from dispersing any funds. The dream of showing up the federal government, of creating a safe haven for stem cell research on California's sunny shores, had taken a decidedly darker and, some say, vengeful turn.

Ortiz, ironically, became one of CIRM's most vocal critics. Just a month after the ballot passed, on the state legislature's first day back in session in Sacramento following the election, she proposed a bill that would change some critical features of Prop. 71 to ensure that the medical and financial benefits of the research would flow back to the citizens of California. When I ask her about why she challenged the very agency she helped to launch, she is very clear that her motives were anything but personal. "Others have put their spin on it," she says, "but I was a big proponent of this. I always have been, and always will be. But I was also keenly aware—did we oversell it? Is it hype? Did it turn out not to be the great promise of cures and therapies we all had hoped it would be? And the answer is a little bit of yes, perhaps. After Prop. 71 passed, I was certainly out there trying to put some oversight and accountability in place."

Ortiz says her primary goal in putting CIRM on the spot was not to derail its mission to fund stem cell research. Rather, it was procedural, to ensure that, as promised during the campaign, the taxpayers funding the work would be able to reap the benefits of their investment, because the state would be entitled to a portion of the profits from any patents or licensing agreements made from the products emerging from the research, and that these products would be available to the state's indigent populations at a reduced, more affordable rate. The initiative itself, she says, made no such guarantees, although it created a body that would address patent sharing and royalty issues. Ortiz wasn't satisfied with the promise or the intentions of the agency in carrying out these promises; she wanted it written into the law. "Look, my legislation was talking about the fact that I recognize that you [CIRM] are self-governing, you get to make your own rules," she says. "But since you get to make your own rules, could you please make some rules that assure that California gets some benefit, and that there's a return to poor people in our overburdened healthcare system?"

Her appeal, she says, was very different from that of the lawsuits, which questioned the legality of having an independent, self-governing entity dispersing taxpayer dollars at arm's length from the state legislature.

And yet despite her protestations to the contrary, some familiar with Ortiz's early efforts in researching the bond issue and pushing through legislation that made stem cell research possible in California find it hard to imagine how she wouldn't feel at least a little betrayed by the way things turned out. "Bob really pissed Deborah off," says Jerry Zucker. "I think that's why she actually started to act against it all after a while.

Bob never gave her credit. He always said it was his brainchild, which is factually not true. It was Deborah Ortiz."

Hall agrees, speculating that if Klein had acknowledged Ortiz in any way, the first few months of CIRM might have been very different. "If Bob had stood with her on the dais after Prop. 71 passed, and said look what we have done, it would have made a difference," he says.

Regardless of Ortiz's motives, from Klein's perspective, her agitating in the legislature was just as detrimental to CIRM and its mission as the lawsuits. "Certainly Bob very quickly came to view her as the enemy," says Hall.

★ ★ ★

Klein describes those first two years of CIRM as difficult but not demoralizing. "It was very, very tough and exhausting, but I really did believe we would get through that," he says. The backers of one of the plaintiffs was the Life Legal Defense Foundation, a pro-life organization that was opposed to embryonic stem cell research. Given the polling that continued to show that the majority of Californians were still in support of taking a leading role in financing stem cell studies, Klein was confident CIRM would prevail.

With the agency's assets frozen until the lawsuits were heard, however (the state was prohibited from floating bonds, because the courts could have invalidated the entire initiative), the institute survived on a start-up loan of three million dollars from the state and an additional five million-dollar gift from Ray and Dagmar Dolby, the founders of Dolby Laboratories—hardly enough for an agency committed to spending an average of $295 million annually to advance stem cell science. To find a way around the legal freeze and to keep CIRM operating, Klein convinced the state to issue an unusual type of bond with an extremely hard sell, asking investors essentially to buy not bonds but bond anticipation notes, another creative financing option in which the buyers willingly accepted the possibility that if the suit against CIRM succeeded in court and the agency was deemed illegal and disbanded, they would not be repaid for their investment. Despite the risk, six private charities poured fourteen million dollars into CIRM's nearly empty coffers.

With the notes sold, by April 2006 CIRM could finally award its first grants—$12.1 million to training programs that would educate and groom the next generation of stem cell researchers at sixteen institutions statewide. But it would take nearly another year before CIRM awarded its first bona fide scientific grants for research experiments, in February 2007. Even then, the seventy-two projects were relatively small, by scientific standards—the largest amounted to eight hundred thousand dollars—and the total, forty-two million, represented roughly the same amount the federal government had doled out for human embryonic stem cell research. The firecracker seemed to have fizzled.

★ ★ ★

While the lawyers and accountants wrestled with the legal and financial operations of the three-year-old institute, CIRM's committee to review scientific grant proposals was raising some intriguing issues of its own. In order to accommodate California's sunshine laws, which required open meetings at publicly funded agencies, CIRM began posting comments on its website from reviewers of the first round of successful grant submissions.

In scientific circles, such airing of the back-and-forth over the merits of proposed experiments is unheard of, almost taboo even after a grant is awarded. Closed-door discussions, scientists argue, are critical to an honest and free exchange of ideas and opinions on key features of scientific experiments, including whether a hypothesis is valid, a methodology is appropriate, or an investigator is qualified to address the scientific issue in question.

Yet here was CIRM, posting comments from one reviewer noting that while the proposal submitted by Hans Keirstead, of UC Irvine, to study stem cells in spinal cord repair was ultimately accepted, it was "scientifically thin," and that Dr. Arnold Kriegstein, director of UCSF's new Eli and Edythe Broad Center of Regenerative Medicine and Stem Cell Research, seemed "naive" about the challenges of testing nerve cells he proposed growing from human embryonic stem cells for use in treatments for epilepsy and Parkinson's disease.

It was a new way of funding science, by popular vote, which brought with it a new way of evaluating science. While it made some uncomfortable, for the most part even the targeted researchers agreed that a little bit of public embarrassment was a small price to pay for finally getting the opportunity to roll up their sleeves and get to work making and studying embryonic stem cells, and turning them into treatments.

And that may ultimately be the final legacy of California's grand experiment—pushing scientists to break free of the familiar way of conducting research and playing by a new set of rules. By its sixth year of doling out grants, CIRM was finding its niche in that most precarious space for researchers and pharmaceutical developers alike, the dreaded valley of death between promising early trials and the more extensive, expensive clinical trials to evaluate a new treatment's potential. It's a no-man's land where venture capitalists are reluctant to tread, and where a therapy isn't proven enough to attract the attention of pharmaceutical companies. But CIRM grants may be just the bridge for carrying academic researchers past this chasm and depositing them on the other side. In 2008 the agency launched its Disease Team Awards, an innovative funding scheme designed to accelerate the process of bringing testable therapies to patients. The concept is to nurture ideas from start to finish, providing a cushion of funding from the lab through early animal testing and into the first safety testing in human patients, thus reducing the likelihood that promising therapies would get left behind for lack of funding. The grants are awarded on a four-year time scale, which motivates researchers to ensure that both the basic science and the studies needed to bring their work to the clinic are completed as quickly as possible.

It's a radical way of funding science, and a departure from the project-by-project approach that defines the NIH. But if it works, it could serve as a model for how science could be done—particularly new science such as stem cell studies. "The experiment of this state to fund stem cell research has way broader implications than we thought," says Weissman of the grants. "They are changing American universities to replace the venture capitalists. It's an investment on the part of CIRM, because stem cell research is a new field. And it needs a new way of thinking. Five years from now, when we have trained all of these stem cell biologists, we could say, 'Is it the state's function to fund only one part of their work? Or do we continue to fund it to its fruition?' If there is a net increase in well-being, health, and money, we will probably keep doing it."

Although the state faced a twenty billion-dollar budget shortfall in 2009, CIRM has been insulated from its growing debt crisis, but that reprieve will last only until the interest payments on the bond sales come due. In 2004, proponents of stem cell research could rely on one compelling argument over all others: They needed CIRM to step in where the government and NIH had stepped out. That, after all, was CIRM's very reason for being. Yet five years after the voters agreed, the federal government began funding embryonic stem cell research again. When CIRM's ten-year budget dries up, will California's voters still feel the same way about stem cells?

Klein is convinced they will. "Our commitment is to patients," he says. "We can move through these models for treatment at a speed that is very difficult for the NIH to duplicate. But at the same time, the basic research of the NIH can help to build and strengthen the underpinnings of the science that we then take and move downstream to patients. Our mission is to get therapies to patients—through small molecule therapies by understanding disease processes better; through human toxicity studies that are more predictive and save massive time and money compared to animal studies; and through cellular therapies that we can move more quickly into clinical trials. If you look structurally at what we can deliver, not only do I think we'll continue to be a major contributor, but I think other states with concentrations of a biotechnology capacity will need to have a mechanism for selling bonds and providing a stable base to really help provide alternative models for innovation in treatments."

It's a grand plan, and one that if not for the success of CIRM to date might be easily dismissed as evidence of more of Klein's posturing.

Cynical observers of CIRM point out that Klein's official tenure ended in 2010—just before the bonds sold to finance the institute in 2004 were due—and the board had not assigned a successor. Instead, it voted to reelect Klein for another six months while a suitable replacement was found.

8

Caught Up in Cloning

As Doug Melton observed CIRM's rocky journey in California, as volatile as it got at times, he was envious of one thing. His group had become the leader in deriving stem cell lines from IVF embryos, yet he remained convinced, as he had from his first days in the field, that he would eventually need to generate stem cells from diabetic patients themselves. In order to truly understand a disease such as diabetes—and exactly why the beta cells stop making insulin—he would need to go to the source. Very few of the embryos he was receiving from the infertility clinics met this requirement; with the donated embryos, he couldn't be choosy and was grateful for what he could get. And while it was true that he hadn't yet succeeded in making any beta cells from even these embryos, he knew that would only be a matter of time. In his mind, it was an issue of when, not if, he or someone else would be successful in generating insulin-making cells from embryonic stem cells. As a scientist, of course, he wouldn't be averse to having his lab reach that milestone first, but as the father of two diabetic children, he was perfectly happy to have someone, *anyone*, make the breakthrough.

The only way to do that, however, was to co-opt the cloning process—the same technique that Wilmut had used to create Dolly, and the same process that had become such ethical and political quicksand for any scientist proposing to use it with human cells. It was the only option available for taking a cell from a diabetic patient and making an embryo that would yield the embryonic stem cells Melton needed. No one, however, had yet succeeded in generating a viable human embryo using nuclear transfer with human cells. But in California, scientists could apply for CIRM grants to at least make the attempt. In Massachusetts, where Melton worked, a state law prevented scientists from creating embryos solely for research purposes, which effectively shut down such efforts for therapeutic cloning.

That didn't mean that others around the country weren't trying— from legitimate scientists, biotech companies, infertility clinics, and even, in one of the oddest and most entertaining chapters in stem cell history, a parade of charlatans and offbeat characters who claimed they were close to generating the first cloned human babies and included a physicist working in Chicago; a pair of infertility specialists from Kentucky and Rome, Italy; and a French cult that believed that all life on earth, including human beings, originated in the laboratory of a beneficent race of aliens. None of which made the scientific case for cloning-based research any easier.

★ ★ ★

It wasn't as if the U.S. government had remained silent on the question of whether to clone or not. Congress simply couldn't agree on the right policy to adopt. While nearly everyone—the American public, legislators on both sides of the abortion debate, and scientists, including Ian Wilmut—agreed that cloning simply to create a genetically identical baby was not only morally unacceptable, but medically irresponsible, there was no consensus on whether any other applications of the cloning technique could be justified.

As Varmus had feared in the days following Wilmut's announcement about Dolly, the two different applications of cloning consistently blurred in the public's mind. While he and others attempted to reiterate that all cloning was not equal, stem cell opponents were more successful in conveying the message that cloning was cloning, no matter the intent.

President Clinton had set the bar in 1997 by establishing the Human

Cloning Prohibition Act, which called for a five-year ban on using nuclear transfer to clone human beings with federal funds, and a voluntary moratorium on the practice using private monies. The proposal was based in large part on the recommendations of the National Bioethics Advisory Commission (NBAC), which Clinton had charged with reviewing human stem cell research. The ban he proposed also forbade cloning to generate embryonic stem cells. Congress, however, never passed that legislation, and neither did it pass several subsequent bills concerning human cloning. The issue instead triggered a years-long debate that remains unresolved today. The United States is one of the few countries that still does not have any legislation regarding human cloning, much less a law that bans the practice to create baby copies.

The stalemate, as always when it came to embryos, revolved around abortion. Pro- and antiabortion legislators tussled over how far to go with nuclear transfer. Those against abortion and stem cell research pushed for a blanket ban on all human cloning activity, while those in favor of embryo research advocated limited use of cloning, specifically to generate stem cells. These advocates felt that an outright ban on all cloning was too restrictive—while no one had successfully generated a human embryo using nuclear transfer yet, they feared that without government support of the science, no one ever would, at least not in the United States. They lobbied for a less inclusive policy that would make the distinction between reproductive and therapeutic cloning.

For the pro-life and antiabortion legislators, however, such a distinction did not go far enough to protect human life. To them, such open-ended regulation represented the first push down a steep and slick slope toward a world in which cloning people joined IVF as yet another reproductive option, and in which human life was reduced to and ultimately devalued as a laboratory process of pick-and-choose, mix-and-match genetic characteristics.

And for a while, it seemed they might be right. Two days after the NBAC announced its recommendations, and even before the commission's chair had an opportunity to officially present the report to President

Clinton, a white-robed French former race car driver sporting a topknot announced that he was launching a human cloning service. Rael, the leader of a worldwide religious cult claiming some thirty-five thousand members who believed aliens had created all human life, offered to clone members of infertile and homosexual couples for "as low as $200,000."

Within months, Richard Seed, a Chicago physicist with a degree from Harvard, joined the fray, announcing that he, too, was attempting to create a human clone—of his wife. "The less everybody knows the better," he cryptically admitted to the *Ottawa Citizen* during a phone interview about his efforts. He would only acknowledge that his wife would also serve as the surrogate for her own clone and carry her to term.

By August 1998, the fever had spread overseas, and Korean infertility researchers claimed to have generated the first human clone via nuclear transfer. The presumably well-meaning scientists failed, however, to follow basic scientific protocol for such a stupendous breakthrough, and not only destroyed their embryo at the four-cell stage, but never wrote up their findings for publication in a scientific journal, making it impossible to verify.

By the spring of 2001, the cloning circus had descended upon Washington, D.C., promising to indeed be the greatest show on earth. The elements were all there, with Representative John Greenwood, a Republican, who called a hearing to debate the pros and cons of human cloning and hash out appropriate legislative regulation of the practice, serving as ringleader. The Raelians fulfilled their purported role as the clowns, since it was so difficult, if not impossible, to take seriously their claim to have both the scientific wherewithal and the funding to sustain a human cloning service. Even the Republicans who had invited them to testify did so somewhat sheepishly. Clearly, as one staffer who was working on the Hill at the time admits, they hoped that the sight of the white-robed cultists would establish, once and for all, how dangerous and ridiculous the whole idea of human cloning was.

To a certain extent, the strategy actually worked. The net result was that all efforts involving the words *human* and *cloning* were tinged with what some bioethicists were beginning to call the "yuck factor." And that left Congress deadlocked over cloning legislation. For Melton, it meant that while he had been successful in drumming up funds for making stem cells from excess IVF embryos, doing the same to make nuclear transfer embryos was an entirely different matter, because it involved cloning.

Melton had no interest in making cloned babies. In fact, he remains opposed to using nuclear transfer solely to produce a clone of a living being. He instead was obsessed with the idea of exploiting the cloning process for its ability to yield stem cells from patients, because it would provide an unprecedented way to watch the cells as they grew, from stem cell to endoderm to pancreatic cell and eventually to an insulin-producing beta cell. It would also be the only potential way to chronicle why that cell loses its ability to produce insulin or respond to glucose.

The uneasiness over creating a life in the lab only to study and destroy it, however, remained a powerful counter to the ultimate goal of helping patients. Even the Howard Hughes Medical Institute, which had been so ready to fund Melton's embryo-based stem cell work, drew the line at therapeutic cloning. So did the JDRF, whose board had supported Melton's previous work because he would be using only embryos destined to be discarded.

Yet again, Melton had no other option than to create what he needed himself. As he had done in deriving new embryonic stem cell lines from IVF embryos, he would simply have to learn to clone human cells himself. Or at least recruit someone who could.

★ ★ ★

If Chad Cowan's initial meeting with Doug Melton was reminiscent of *Love Story*, then Kevin Eggan's was more like *The Godfather*.

A native of Illinois, Eggan still exudes a Midwestern gregariousness—he's not averse to chatting up seatmates on plane rides—and possesses dark-haired good looks that once earned him an invitation from *GQ* to pose for a feature on rock stars in science (which he declined). His appearance, however, shouldn't detract from a keen intellect that has earned him a MacArthur "genius" grant and the distinction of being one of the youngest junior faculty members accepted at Harvard.

By 2002, with a Ph.D. in molecular biology, Eggan was establishing himself as one of the country's leading experts in the nuclear transfer technique—and just the eager, talented, up-and-coming researcher Melton was seeking. Eggan's education as a scientist had occurred in the formidable shadow of cloning, and he, like many of his colleagues earning their doctorates at the time, could not help but be swept away by its scientific possibilities. As he was completing his Ph.D. at MIT in 1996, Wilmut cloned Dolly, and in the following year, using the same procedure, a group at the University of Hawai'i cloned a mouse, Cumulina, for the first time. And while Cumulina hardly received the same media attention that Dolly did, for students of science like Eggan, the mouse's birth was potentially more momentous because, as he says, "sheep and cows as cloning model organisms were not exactly convenient as molecular genetic models."

Like Melton, Eggan appreciated the power that Cumulina would have for studying what was still the black box of development. Unlike his diabetes-focused professor, the young scientist had yet to fully appreciate the power that cloning could have on studying human disease. But he was sharp enough to sense a once-in-a-lifetime chance not only to witness but to shape an emerging field of science. "This seemed like an incredible new development in developmental biology, and I didn't want to miss out on this opportunity," he says. "So I jumped to Rudolph's lab."

And that's where Melton found him. Rudolph Jaenisch was by then quite familiar with mouse embryonic stem cells, having worked with them for the almost two decades that they had been around. Jaenisch was making a specialty of understanding the details of how cloning worked, and was trying to recapitulate the steps that an adult cell would have to take in order to reprogram itself as an entirely new animal. As part of those studies, he assigned Eggan the task of learning and perfecting nuclear transfer in mice—which he did, and with such skill that he was ultimately brought to Melton's attention.

An entertaining storyteller, Eggan recounts how Melton, whom he had never met and knew only by reputation, recruited him from MIT to Harvard. It's hardly the traditional tale of nerve-wracking interviews and probing reviews. In fact, he says, it unfolded more like his own personal *Godfather* saga.

"I was walking around the lab one day, and this Israeli postdoc in Rudolph's lab walks up to me and says, 'Hey Kevin, my Israeli friend in Doug's lab says Doug wants to meet you,'" he begins. "I barely even knew who [he] was. Remember that Doug is pretty reclusive—you almost have to seek him out. Scientifically, I always imagined him sitting in the back of an Italian restaurant, if you know what I mean," he says, a smile spreading across his face. "So I say to my friend, 'Well, you tell your friend in Doug's lab that I'd like to meet Doug.'

"It goes on like this, and a week later, this guy comes up to me and says, 'My friend in Doug's lab says that Doug wants you to come to give a seminar.' I'm like, 'Okay, you tell your friend in Doug's lab—and this is getting weird—that I'll come and give a seminar.' So the whole thing got arranged through this Israeli arbitration without any direct contact between Doug and me."

It was late fall in 2002, and a couple of months later, Eggan hopped on his bike for the chilly fifteen-minute ride along the Charles River from MIT's Kendall Square to Melton's lab, just off of Harvard Square. After making his presentation, the two moved to Melton's office. "That was the first time I ever heard this simple idea articulated. He said, 'Why not use what you do, and what I can do now with human embryonic stem cells, to make a limitless supply of cells that get sick in people? And surely we could figure out what was going on with diseases like diabetes if we could do that.'" Eggan pauses to emphasize the impact this idea, so familiar now but still only a seed in Melton's mind in 2002, had on his thought processes. Until then, he says, the nuclear transfer technique he was learning was to him just that—a technique and a means to better understand how cells reprogram themselves back to the embryonic state. Cloning was merely a tool to answer scientific questions for science's sake. But with his first exposure to Melton's singular focus on finding a cure for diabetes, he suddenly saw the potential for applying therapeutic cloning to treat and cure disease. "Put the details aside, and think about that for a second," he says. "I was sold. I drank the Kool-Aid immediately."

And for Eggan, it was a powerful elixir. After spending nearly four years perfecting the cloning technique with mouse cells, he was itching to try it with human cells.

But based on his trial run with the mice, he knew it was going to be far from easy.

★ ★ ★

When Jaenisch presented Eggan with the task of learning and setting up nuclear transfer in his lab at MIT, Eggan's only credentials for performing the delicate micromanipulation work it required were, he freely admits, rather unscientific. "Super Mario Brothers," he says simply about the extent of his skills in manipulating microscopes and handling embryos. So it wasn't a total surprise that the first year and a half of attempts to clone mice, he concedes, "were not wonderful. It was actually quite traumatic."

Aware of the difficulties, Jaenisch dispatched Eggan to Hawaii to learn nuances of the technique that only those who have tried it could teach. By this time, the island scientists had succeeded in breeding several generations of cloned mice pups after Cumulina's birth.

Arriving in Honolulu, however, Eggan wasn't prepared for what he saw. The laboratory, such as it was, was nothing like the modern, well-stocked hubs of high technology he was accustomed to back at MIT. It occupied an unused part of a windowless cinder-block building that served as the loading dock for the university's food services. The researchers kept their mice in an old walk-in freezer that had been shut off. And since the facility had not been built with high-level science in mind, it lacked the vibration-buffering lab benches that are the staple of any science department in any university. Instead, students conducted experiments on sturdy, 1950s-era desks, whose legs were cushioned by tennis balls that had been split open and slipped onto the ends—a trick borrowed from the elderly who do the same to stabilize their walkers. The balls were an attempt to keep the desks from moving too much when anyone performed delicate procedures such as puncturing tiny mouse eggs and squirting them with sperm from a pipette under a high-powered microscope.

"I was like, 'Whoa,'" Eggan recalls about his initial impressions of the Aloha lab. "I thought, 'I'm going to do tissue culture, and embryonic stem cell culture, just on a desk basically.'" He pauses, shaking his head. "Which worked remarkably well."

As primitive as the conditions seemed to Eggan, they were certainly good enough to lead the lab, run by an energetic and engaging Japanese scientist named Ryuzo Yanagimachi, to some important biological firsts. Yanagimachi, or Yana, as he is known, is credited with improving the success rate of pregnancy with IVF by injecting the swimmers into eggs with a more refined and targeted piezo pipette.

It turned out that same technique would set the team up for success in cloning as well. Although, truth be told, it wasn't exactly a project that Yana had initially wanted his lab to pursue. The milestone experiment that became Cumulina actually came about because Yana's star postdoctoral student couldn't swim.

When he arrived in Honolulu from Japan, Teruhito Wakayama, a shy, serious man, never imagined he would gain such fame in cloning. During our first meeting in New York, shortly after his cloning breakthrough in 1998, he is apologetic about his English, although it is perfectly acceptable, and reluctant to make too much of his work. In Yana's lab, his first project was to break down the act of fertilization on a molecular level to figure out how sperm excites the egg enough to prompt it to divide and grow into an embryo. It turns out that the mystery factor is conserved across disparate species—you can inject sea urchin sperm into a mouse egg and the egg will start to divide as if it had been fertilized by a mouse sperm. The embryo will eventually die, but for a few days, it will act as if it were on its way to becoming a mouse pup.

So that's exactly where Wakayama focused his attention—on sea urchin sperm. The lab's location off some of the richest coastal waters in the islands had one distinct advantage. "The cost was free," he tells me of his urchin resource during a recent talk in his office in Kobe, Japan, where he now heads a lab at the RIKEN Center for Developmental Biology. The only problem was that he had to collect the spiny creatures himself, several times a week. To anyone enamored of the sun and surf and sea, it would have been the perfect scenario—visit the beach almost daily, go for a swim to collect sea urchins, and spend the remainder of the day back in the lab doing experiments. Ideal, that is, if you don't mind the swim. Wakayama did. So instead of a welcome respite from the tedium of lab work, collecting the sea creatures became a bit of a chore. "I must go to the ocean from the university, and I cannot swim," he explains.

Just next door to the lab, in that converted freezer, however, lived another species that was much more accessible—mice. So he decided to switch. They were more expensive to work with, but at least collecting them didn't require a chilling and intimidating dip in the ocean.

Having read about Ian Wilmut's success with Dolly, Wakayama was inspired one day to try something similar with his mice. In collecting the eggs from females that had ovulated, he often also swiped up something called cumulus cells, which, as their name implies, surround oocytes like a cloud of nourishing tissue. Unlike the egg and sperm, which contain only half the full complement of chromosomes, cumulus cells are fully developed adult cells with two copies of every chromosome. Following Wilmut's lead with Dolly, Wakayama decided to squeeze out the nucleus of one of his mouse eggs and replace it with the nucleus of a cumulus cell from another female donor—just to see what would happen. To his surprise, the experiment worked and in eighteen days, he saw a litter of mouse pups.

"I was very surprised," he tells me. "I believed mouse cloning was impossible because many papers said mouse was one of the most difficult species to clone. I believed the papers and the textbooks."

If he was surprised, Yana was even more so when he learned of the experiment and its success at the same time. "I did the experiment without telling him," admits Wakayama with a laugh. "He didn't know until the success."

It's not the best way for a postdoc to endear himself to his mentor, but in Wakayama's case, it worked. And it put Yana's lab on the short list of "it" laboratories in the hot new field of stem cell research.

★ ★ ★

Eggan is still amused when he recalls his first days in Honolulu and the incongruity of seeing highly sophisticated work performed in a laboratory that was a throwback to a more primitive time. "The building was ambient to the outside, so it was eighty degrees in the lab," he says. "It was just a bunch of Japanese and American guys micromanipulating with no shirts on. It was a really weird environment."

The collaboration, though, was a mutually beneficial one. As Yana's group continued to clone more litters of mice, they began noticing differences, beginning with the pregnancies, between the pups born via the nuclear transfer process and those gestated via the conventional IVF technique. The cloning procedure produced a high abortion rate, and if the pregnancies went to term, the pups tended to be large and overweight. Another Japanese postdoc in Yana's lab, Hideo Akutsu, began wondering whether something in the reprogramming of the adult cells was going awry. Jaenisch was the perfect person to ask. While the Hawaiians had perfected the cloning technique, Jaenisch was making headway in exposing the molecular mysteries of reprogramming—how a mature, adult cell turned back the clock and rewrote its developmental history. Eggan would learn the nuclear transfer process from Akutsu, and Akutsu would come to Massachusetts to pick up more knowledge about the unknowns of molecular reprogramming.

For Eggan, it wasn't a bad way to spend a summer. "The day would start at seven o'clock, and because of the hormonal cycles of mice, we'd basically be done by three," he recalls. "I would get on Hideo's bike, I would ride down to Waikiki, I would surf until sunset, I would go back to my hotel room, pass out, and the next day we'd do it all again." He grins. "It was pretty good."

When he tried to apply what he learned back in Boston, however, things didn't appear so sanguine. No matter what he tried, the cells didn't seem to "take," and Eggan failed repeatedly to get adult mouse cells to successfully seed an oocyte and generate stem cells. He made several trips to Honolulu, carting his culture media and a kitchen sink of chemicals and growth factors and enzymes, hoping that he and Akutsu, with whom he had become good friends, could figure out what was wrong with his formula.

The failures, says Akutsu, made Eggan singularly obsessive about mastering the technique. "Kevin was very, very aggressive about learning, very curious about the microinjection technique," he says.

But as many times as he tried, Eggan simply couldn't get the cells to grow as they did in Honolulu. Nuclear transfer is a finicky process to begin with—even Wakayama and the Hawaiian team were only able to reach a peak efficiency of 2 to 3 percent out of their manipulated embryos. "There was a year and a half where there was utter failure at MIT," Eggan says. Day after day he would sit down at his microscope and hope to see the eggs he had injected with a mouse skin cell begin to divide. But day after day he was disappointed. The embryos almost seemed to be taunting him; he could see the nucleus of the injected cell inside the egg, so he knew that the development process was turned on, but somehow time after time it stalled at the first step.

Frustrated, and concerned that it was his technique that was at fault, he set up control experiments, performing exactly the same steps in Honolulu and in Boston, and still the cells would grow in Hawaii but fail to thrive in Massachusetts. "So I knew it wasn't me," he says. Desperate, he even tried to mimic the steamy temperatures in the island lab in his tissue culture room in the dead of the Boston winter. "It was eighty degrees in my micromanipulation room at MIT, because we tried every-thing; I didn't know what was wrong. I was close to the no-go point I'd agreed on for myself, that I was going to do something else if I couldn't get it to work."

The culprit, it turned out, was a single reagent that Wakayama had used to jump-start the sperm. Not many biologists used it, and not many would have thought to use it, says Eggan, but it was part of Wakayama's magical mix. In fact he didn't think much of it and never thought it was something that might be critical to his success. For Eggan, after 450 days of failure, in the end it was a single chemical that made the difference between his attempts in Honolulu and those in Boston. "As soon as we fixed that, everything worked," he says.

But even with the Hawaiian "recipe" finally complete, the nuclear transfer technique was still more art than science. In the same way that handmade crafts are similar but not identical, the stem cell lines pro-duced by nuclear transfer differ slightly depending upon whose hands transfer the donated nucleus and then nurture the stem cell along.

Eggan became quite adept at the technique, and was the first in the continental United States to generate a mouse clone. But his ultimate goal, as inspired by Melton, was to apply the technology to human cells. So in 2004, he was finally ready to try his hand at starting the whole process of trial and error all over again, this time with human cells. And Melton was promising to make it happen. "I wanted to do human nuclear transfer; that's the reason I went to Doug's lab," Eggan says. "Doug convinced me I would be able to convince the university to let me do it."

★ ★ ★

And ultimately, the university did. But it would take two years and the intensive scrutiny of eight different review committees before the exper-iments were given the green light. While their goal was not to create human clones, the experiments that Eggan and Melton were proposing were, in essence, cloning experiments. In order to access stem cells, they would have to clone human embryos and then destroy them.

This strategy now separated them ethically from using excess IVF embryos to generate embryonic stem cell lines, as Melton had been doing. Pro-life groups, leaders in the Catholic Church, and antiabortion advocates strongly opposed the idea of any such study of human embryos, adamantly maintaining that life begins at conception and therefore such studies devalued the sanctity of human life.

Eggan and Melton, along with scientists such as John Gurdon, began to draw their own ethical lines around these embryos, arguing that such therapeutic cloning really couldn't—and shouldn't—be perceived as cloning at all, and that the four- to five-day-old embryos they created did not constitute life in the way pro-life supporters were maintaining. Drawing on biological criteria for defining potentiality, these researchers believed that until a human embryo is implanted into a womb, it has no possibility of becoming a human life, and therefore cannot be considered to have the same value and rights of an implanted embryo.

"What people usually don't remember or don't choose to remember is that the human embryo is in no way a potential human being, because it hasn't been implanted," Gurdon says. "In the absence of implantation, there is no conceivable way that an embryo would ever become a human being. One could well argue that the preimplantation human embryo has not at all the potential of being human."

That hardly meant, however, that the stem cell biologists working with them were any less respectful of human life or too callous to appreciate the nature of the material they were handling. Within Eggan's own lab, for instance, researchers were beginning to recognize just how ethically complex their roles as stem cell scientists were.

As one of Eggan's postdoctoral fellows, John Dimos was busy making new stem cell lines from diseased IVF embryos in late 2007 when he received a call to pick up some donated embryos from a nearby IVF clinic. Dimos came to Eggan's lab precisely because of Eggan's interest in translating stem cell technology into new treatments for disease; he was aware of Eggan's efforts to apply nuclear transfer to human cells, and was eager to participate in those studies. In the meantime, he took advantage of creating as many lines from donated embryos with known diseases as he could. Most of these conditions belonged to the dozen or so single-gene disorders that could be detected with preimplantation genetic diagnosis of IVF embryos—conditions such as cystic fibrosis, Duchenne muscular dystrophy, fragile X syndrome, and sickle cell anemia. Technicians removed one cell of a four- to ten-cell blastula (which forms about three to four days after fertilization in a dish) and analyzed its chromosomes for specific mutations. If the embryo tested positive, it was not implanted; if the parents consented, it was either destroyed or donated for research. On that day, Dimos was told that one of the clinics working with the Harvard labs had embryos with the genetic mutation for Huntington's. He agreed to pick them up and took the cab over to Lexington, Massachusetts.

"When I arrived, I met with the clinician. She was extraordinarily upset," he tells me. "Evidently she knew this couple, and they had Huntington's embryos and she couldn't say anything. She was visibly shaking as she transferred the embryos in their little bar-coded tubes to me." While Huntington's disease can be detected through genetic testing, the screen has launched a series of difficult ethical questions as people at risk for the disease struggle with whether or not to get tested. There is no cure, so not everyone is ready to learn if they harbor the Huntington's gene. The couple from whom the embryos came had decided they did not want to know if they were carriers. But because their embryos were affected by the disease, either the mother or the father, or possibly even both, had the genetic abnormality. To respect their wishes, they would not be told that their embryos had tested positive for Huntington's; they would only be informed that their IVF cycle had not succeeded.

The encounter was a reminder to Dimos of what was at stake with every procedure he conducted under the microscope. It wasn't, he says, as if he had suddenly become uncomfortable with his chosen profession. It was more that his studies took on an added significance that they had been lacking before. "I'd be lying if I said I cared any differently or was motivated any differently," he says, choosing his words carefully. "But it felt different." When he returned to the lab to extract the stem cells from those embryos, he says, "it was the first time that I wasn't just getting stem cells. It was that I was destroying a human embryo to get stem cells, and the first time I had any emotion associated with it. I saw that there were much broader implications of the work—or at least that there ought to be."

★ ★ ★

Melton wasn't oblivious to the moral implications of his work either. Paying homage to his degree in history of science, he immersed himself in the thorny ethical questions his work raised by teaching a course to undergraduates on the ethics of stem cell science with Michael Sandel. He and Sandel, a professor of political philosophy who's well known for a popular undergraduate course titled Justice, laid out for students scenarios that would cut through political rhetoric and obfuscation to expose what he felt were people's core beliefs about the status of an embryo.

In one of his favorite examples, he asked students to consider what would happen if a fire erupted in a fertility clinic: If they had time to save only one victim and had to choose between a six-year-old child and a tray of embryos, which would they choose? The class also gave Melton the opportunity to interact with leading opponents of stem cell research— among them Richard Doerflinger, of the United States Conference of Catholic Bishops, and Leon Kass, head of President Bush's President's Council on Bioethics, who were invited to present the critical side of Melton's work. During Doerflinger's session, Melton famously asked him whether, if Melton were successful in using human embryonic stem cells to develop a treatment to cure a disease affecting one of Doerflinger's children, he would refuse to use it on moral grounds. "I said that I would think as a matter of principle I would want to be able to say no," Doerflinger tells me. "But I hope in the meantime I would be working for a world in which that [cure] was not the only way we have to deal with those conditions." The class, both Melton and Doerflinger recall, booed.

Such exercises, Melton believes, get at the core of how we define our moral philosophy. Taking the stem cell debate out of the political arena, for example, stripping away the loaded meanings of words such as *embryo* and *life*, and laying it bare as a choice about how you would make decisions about your own children, he says, "really speaks to the core of what you believe. Because if something is wrong, then it's wrong. You wouldn't do it."

In discussions with Sandel, Melton came to the same conclusion that the Human Embryo Research Panel had in 1994—that things in nature, including living things, can be respected at varying levels, but without necessarily being accorded equal moral status. Sandel often uses the analogy of an acorn, which is a potential tree, but notes that acorns aren't treated in the same way as old-growth forests are. In his view, the fact that all people were once blastocysts does not mean that all blastocysts are people.

The Harvard Institutional Review Board (IRB) eventually agreed with this reasoning, and acknowledged in 2006 that the research Eggan and Melton proposed—to make human embryos via nuclear transfer— was justified and, on ethical grounds, reasonable. The university's program was the first in the nation to ask women to voluntarily donate eggs to be used purely for research purposes. (Since the 1980s, women had been giving eggs to help infertile couples have children, but no academic scientists had ever asked them to willingly undergo the invasive and potentially dangerous procedure of hyperstimulating their ovaries to release eggs and then surgically removing them with no intent of creating a baby.) "We certainly realize that all eyes are upon us," Eggan noted on the day the program was announced.

Across the country, at UCSF, Dr. Renee Reijo Pera, then codirector of the university's Human Embryonic Stem Cell Research Center, began a similar project. What Pera (now at Stanford), Eggan, and Melton realized was that without the ability to generate stem cell lines from patients, the entire field of stem cell science would be hampered in a significant way. Even the Harvard program, which had become one of the nation's largest in collecting excess IVF embryos to make stem cells from them, included only a few embryos with known diseases. And even those embryos, just by virtue of the types of couples taking advantage of IVF services, were more likely to harbor uncommon genetic diseases

such as sickle cell anemia and muscular dystrophy than the more preva-
lent conditions of diabetes, Parkinson's, and heart disease.

What Eggan and Melton had going for them was a forward-thinking
university administration led by Larry Summers, who was aggressively
promoting the life sciences in general and was particularly eager to establish Harvard as a premier institute for stem cell science in the absence of
significant government support of the field. Yet the university was still
a scientific island in a politically hostile ocean of opposition. Thanks to
stem cell critics who had equated therapeutic cloning with reproductive cloning and drilled home the dangers of the practice, no government group had produced recommendations favoring nuclear transfer
with human cells, much less guidelines or protocols for how to establish
and operate a program. The last national ethics group to support therapeutic cloning was the 1994 Human Embryo Research Panel; the
National Bioethics Advisory Commission in 1999 and the President's
Council on Bioethics under the Bush administration both advised
against federal support of the studies. Legislatively, Dickey-Wicker was
in place, forbidding the harm or destruction of any embryos studied with
federal funds.

So working with Harvard's IRB, Melton and Eggan had to devise
a pioneering protocol for recruiting egg donors, building upon their
existing relationship with Boston IVF. At the review board's recommendation, these women would be recruited through newspaper and
other advertisements, and not through Boston IVF's clinics, in order to
avoid any whiff of coercion. But the biggest challenge for the project,
and by far the issue that engaged reviewers the most, was whether the
women should be paid. Because there were no such programs anywhere
else in the nation, the Harvard effort would be setting a precedent for
whether egg donation for stem cell research would be any different from
other types of studies or whether they warranted consideration above
and beyond accepted ethical practices. Was the act of giving eggs for
stem cell research, for example, similar to oocyte donation for reproductive purposes, for which women were paid, or was there something

special about egg donation for stem cell research that required some type of additional protection?

Ultimately, the Harvard IRB decided egg donation for stem cell research was no different from voluntarily providing eggs for reproductive purposes, and elected to compensate women for their efforts as IVF clinics do. It was a position that Eggan supported. "And then," he says, "everyone else got involved."

"Everyone" included the institutes and organizations that until then had been silent on the ethics of therapeutic cloning because of its contentiousness. But because Melton planned to use the basement lab created with HHMI money, he informed Thomas Cech of his intention to also use it as the base of operations for the more controversial therapeutic cloning.

Cech assembled the institute's bioethics panel, which consisted primarily of the same people, led by Laurie Zoloth, who had advised him on Melton's previous request to create human ES cell lines from excess IVF embryos. This time, however, the ethical stakes were higher. "They were the first people ever to consider egg donation for SCNT (somatic cell nuclear transfer) research, and they got it wrong, I think," Eggan says, his strong feelings on the subject becoming obvious. "They were well-minded bioethicists, but I don't feel like they were well steeped in human subjects research. They basically looked at this from a first-principle perspective. They reached a conclusion that I think was overly paternalistic."

Eggan and Zoloth continue to amicably disagree on the panel's advice, which was not to compensate women for donating eggs for therapeutic cloning studies. "Some things need to be outside the marketplace," Zoloth says in explaining the panel's rationale. She points to organ donation and aspects of human sexuality that we as a society deem immoral and illegal to sell for a fee, and makes the case that human eggs should fall into the same category. "We don't buy body parts; we don't do certain things because we feel some things ought not to bought and sold," she says, "because the marketplace economy is a dreadful way to think about justice."

Eggan believes otherwise, pointing to a hypocrisy in the reasoning that women would be enticed by money to donate eggs for reproductive purposes, not to mention clinical trials of drugs and other therapies of all kinds. "We pay people in the U.S. to participate in human subjects trials that are much more dangerous than egg donation," he says, "and they don't benefit at all." (Zoloth notes that the policies surrounding egg and sperm donation for IVF occurred largely "without the intervention of bioethicists. We were out of it; they didn't ask us at all," since government restrictions on embryo research drove much of the infertility research into the private sector.)

In Eggan's view, stem cell science was beginning to earn a "special" status in the minds of the public and even among bioethicists and legislators, which wasn't doing the field any favors. Owing to the prime-time attention Bush had given the field with his broadcasted executive order, and the endless debates on Capitol Hill, there was now embedded into any stem cell discussion a sense of what bioethicist Insoo Hyun, at Case Western Reserve University, calls "stem cell exceptionalism."

That perception filtered into the egg donation guidelines that the National Academy of Sciences was drafting, as it, too, opposed paying women donating to stem cell studies. It wasn't long before California, then Missouri, and even Massachusetts, home of the Harvard Stem Cell Institute, passed laws opposing compensation. "The idea was that [the stem cell field] was already being assailed from the right on the question of the moral status of the embryo, so we couldn't afford to be assailed from the left over these issues of women's reproductive health," says Eggan.

The payment issue proved to be the death knell for Harvard's project. In the end, despite a flurry of initial interest from potential volunteers, only one woman agreed to participate before Melton and Eggan decided to terminate the program after nearly a year.

★ ★ ★

Off campus, things were not much better. In Washington, the pressure on embryonic stem cell research, and in particular on nuclear transfer, continued to mount. By 2006, representatives Diana DeGette of Colorado and Mike Castle of Delaware led efforts in Congress to expand federal support of embryonic stem cell research, managing to gain some traction in both the House and the Senate. Staunch Republicans and antiabortion politicians, including Bill Frist and Orrin Hatch in the Senate, joined the chorus in support of stem cell research and helped pass more expansive legislation for the field twice. But both times, even these compelling advocates failed to generate the two-thirds majority needed to override a presidential veto. As he had promised in 2001, Bush promptly exercised his right to shut down both bills when they reached the White House.

It was clear that any research with stem cells, not to mention attempts at therapeutic cloning to get at them, was still a nonstarter in the United States.

The President's Council on Bioethics, a group of experts handpicked by Bush to study stem cells, among other biomedical issues, was busy churning out reports on the implications and applications of stem cell science. But the reviews had the unmistakable whiff of the stem cell–wary White House. Headed by Leon Kass, a molecular biologist who had turned his energies toward considering the moral and social questions raised by science, the council released white papers on human cloning, stem cell research, and alternative sources of stem cells. Kass was well known for his wariness of the unimpeded march of science, calling instead for what he termed the "wisdom of repugnance," a trusting of the visceral, intuitive sense of when something is ethically and morally wrong. "Repugnance is the emotional expression of deep wisdom," he wrote in a *New Republic* essay in 1997, in response to Dolly's birth, "beyond reason's power fully to articulate it." The council's second report on stem cells, which focused on methods for obtaining stem cells that did not involve embryos, was meant, Kass told reporters at the time, as a provocative invitation for researchers to think "creatively" about ways both to satisfy the scientific potential of stem cells and to resolve the moral dilemmas.

But regardless of how neutral and open-ended the council's language was, the report only salted the wounds of the scientific community, which was still smarting over the beating its stem cell members had received when Bush restricted funding of their work. The council outlined four other ways for obtaining pluripotent stem cells that the members argued would exploit the promise of stem cells without destroying embryos: 1) extracting the stem cells from dead embryos; 2) growing them from single-cell biopsies taken from embryos; 3) harvesting them from an artificial, embryo-like "cellular system" that lacked crucial genes and was thus prevented from growing into a proper embryo; and 4) reprogramming adult cells back to their embryonic state directly, without passing through an embryo phase.

From both scientific and ethical perspectives, the first two methods were relatively straightforward, since they had precedent: Organ donation involved the use of tissues and organs from the recently deceased, and preimplantation genetic diagnosis was routinely used in IVF clinics, with thousands of healthy babies born from the embryos that had been biopsied. The last option was, at the time, a scientific fantasy, but represented the holy grail: being able to simply turn back the clock on a mature cell and bypassing the need for eggs and embryos altogether.

Which left the third option, one championed by council member Dr. William Hurlbut, a physician and bioethicist from Stanford who favored the pro-life position that life begins at conception. Hurlbut's proposal involved tweaking nuclear transfer so that an embryo, as he defined it, was not created at all. What resulted instead was something resembling an embryo, which was never designed to develop into a viable human being. What altered nuclear transfer (ANT) created, he maintained, was not a defective embryo but "a laboratory construct, an artifact that fails to form with any contiguity and integrity." In fact, he argued, it was closer to a teratoma—the disorganized mass of wildly and randomly growing cells from various tissues in the body—than it was to an embryo.

Hurlbut continues to believe that ANT is a reasonable alternative to obtaining stem cells, even after breakthroughs in reprogramming cells have made the last choice more realistic. And he maintains his stringent definition of what an embryo is, a definition that he acknowledges aligns pretty closely to that of pro-life defenders. "What characterizes an embryo is that it has within itself imminent powers, indwelling powers for its own development," he explains. "It needs context for protection and nurture, but it has the program for its own development in it. Unlike an assembly of parts, living things begin as whole living organisms, and then produce their parts. In biology, the whole precedes, then produces parts. Which is why a one-celled embryo is a living human organism."

Hurlbut's position on stem cells emanates from this strongly held belief in the equal and unchanging moral status of the embryo, from the moment of conception onward. ANT, he says, "is a method of creating a lineage of cells that has characteristics of tissue culture but not the integrated unity and diversity of parts that are the defining character of an embryo, an organism."

Biologically, at least, ANT is feasible. Seven months after Hurlbut's white paper describing the process was released in 2004, Rudolph Jaenisch, at MIT, put the idea to the test. He generated a series of mice that lacked a gene critical to the development of the trophectoderm, the precursor to the placenta. Without a proper trophectoderm, an embryo is unable to implant in the uterus and fails to develop further. After introducing a defect in the primary gene responsible using tail cells of mice, Jaenisch used these genetically modified cells as donors in the nuclear transfer process. The resulting blastocysts, as expected, failed to develop a proper trophectoderm, and were by many measures abnormal. They did, however, produce an inner cell mass of embryonic stem cells that for all intents and purposes were as pluripotent as stem cells derived from normal mouse tail cells.

Even Jaenisch, however, admitted that the process was problematic and stressed there was still no evidence that the same gene, or even others, would work the same way in generating human versions of the hampered blastocysts. Plus, he had to use a retrovirus to bully the defective gene into the cells, and such agents could prompt the cell to turn cancerous.

To stem cell scientists like Melton, ANT represented yet another round of the word-hockey playing out in the press, on the congressional floor, and in the White House over the definitions of *embryo, life,* and *moral status.* And, from both a scientific and an ethical point of view, Melton couldn't see how creating such an intentionally manipulated "embryo-like" entity would necessarily address the moral concerns of those who were bothered by nuclear transfer. "An embryo is an embryo is an embryo," he told me at one point in our discussions of ANT. Commenting on the proposal in the *New England Journal of Medicine,* he and colleagues at the Harvard Stem Cell Institute described ANT as "flawed." The argument that a defect introduced into an embryo renders it no longer an embryo seemed a semantic biological sleight of hand to Melton, who countered that you could also consider such an embryo "normal" until the defective gene was triggered and then killed the embryo. Why, argued Melton, would you go through the machinations of messing up an embryo when it would be much easier—and, in his view, ultimately more ethically defensible—to simply do nuclear transfer the straightforward way?

★ ★ ★

Driving Melton's obstinate push for therapeutic cloning with human cells were the first tantalizing hints that if he or others could get it to work, the rewards, as far as exposing the cell-based machinations of disease, would be tremendous. In 2002, Jaenisch and George Daley, then a postdoc in the lab, who would go on to study blood stem cells at Children's Hospital Boston and join the Harvard Stem Cell Institute, proved that therapeutic cloning could actually cure a disease. Working with mice suffering from an immune disorder caused by a mutated gene, they cloned cells from diseased animals using nuclear transfer, just as Yanagimachi's team in Hawaii had done with their mice, and then removed the embryonic stem cells, fixing the defective gene before reintroducing the now healthy cells into the sick animals. Immediately the mice started making healthy versions of the new immune cells.

The chasm separating mouse from man, however, was too great. And with such a poor response to their human program for therapeutic cloning, Eggan was forced to find stand-ins for the human eggs. What he needed was something that contained as much of the egg environment as possible—its unique cocktail of growth factors, hormones, enzymes, and other proteins that made the embryonic environment a hotbed of developmental potential. Figuring that an embryonic stem cell, newly hatched from such an environment, would still retain some of this critical mélange, he fused a human adult cell with a human embryonic stem cell from one of the lines Melton's lab had created. Indeed, the stem cell was able to successfully silence the DNA of the adult cell and essentially reprogram it to give rise to all three of the germ layers—ectoderm, mesoderm, and endoderm. But as with many an understudy, the substitute fell short of the real thing. Because all of its reprogramming abilities were encoded in its genes, the resulting cells had double their complement of chromosomes—one set from the adult cell and another from the stem cell. In the classic nuclear transfer experiment, the egg's DNA was removed, leaving the resulting stem cells with the proper single set of chromosomes. But in the fusion version, removing the DNA would eliminate the very source of reprogramming power, so it was left in, which meant that the subsequent cells possessed too many chromosomes and could never be transplanted into patients as a potential cell therapy to treat disease.

It was a valiant effort, but one that Eggan knew was only a stopgap. Using the one donor's eggs, he had not been able to coax out a line of embryonic stem cells.

He couldn't perform the first nuclear transfer experiments with human cells. But someone else soon would.

9

The Rise and Fall of a "Supreme Scientist"

It was often still dark outside when Woo Suk Hwang left his home just outside South Korea's capital city of Seoul. In half an hour, his government-appointed driver sped him through the city's normally congested streets, still deserted at five o'clock, and deposited him at Building 85 on the campus of Seoul National University. The building, too, was nearly unoccupied as he made his way from the basement garage to his sixth-floor labs. The only people he might have encountered were members of his own team, bleary-eyed, exhausted after a night's work tending to the cells that had become a matter of national pride not only for Hwang's lab but for the Korean people as a whole, even those who had never stepped foot in a science lab and had no idea what a stem cell was. By six A.M., he had donned a blue paper jumpsuit that covered him from neck to ankles, a matching bonnet that encased the neatly combed black hair just beginning to show the first sprinkles of gray, paper slippers, and a mask.

Like an astronaut preparing to enter a spacecraft, Hwang stepped through a series of doors into progressively sterile rooms until he encountered a closet-like chamber equipped with glass doors on either side. As he walked in and locked himself inside, gusts of air pummeled him from all directions and whisked away any stray lint, hairs, or pathogens that may have hitched a ride on his disposable blue suit. Exiting the air lock, he entered the inner sanctum, where the world's first custom-made human embryonic stem cells awaited. Their home was a state-of-the art incubator no larger than a mini-refrigerator. Hwang pulled it open, selecting one of the dozens of clear plastic dishes neatly stacked on shelves inside, and turned to a nearby microscope. Within minutes, he was scrutinizing the world-renowned cells. It was a routine he performed every day of the year when he was not traveling.

It took Hwang about an hour to complete his regular scan of the thousands of cells. But exactly what he was actually seeing at the other end of the microscope during those daily dawn reviews throughout 2004 and 2005 still isn't clear. Within a year, those cells would become the focus of a national investigation and a university inquiry, and the source of one of the most mystifying medical scandals in recent history.

★ ★ ★

I first met Woo Suk Hwang in the spring of 2005. A veterinary scientist by training, he had by then already achieved rock star status in Korea; globally, he was the most sought-after researcher in the stem cell field. In 2004, he had dropped, seemingly from the scientific ether, onto the front pages of the world's newspapers when he claimed to have successfully grown a line of stem cells from a cloned human embryo. In two short years, he would repeat that milestone by turning the first patient cells into a line of embryonic stem cells.

His feats came nearly a decade after Dolly, and over six years following Jamie Thomson's and John Gearhart's creation of the first human embryonic stem cells. In that time, scientists had been throwing themselves into a roller derby race to marry the two achievements for another type of breakthrough—generating embryonic stem cells from an adult human's skin cell. Since Dolly's birth, the list of animals that had been cloned using nuclear transfer was long and growing—mice, cats, horses, cows, rabbits, goats, and even rare species like the gaur. And since Thomson had first nurtured his human embryonic stem cells from frozen IVF embryos, handfuls of researchers had also been able to generate their own. But none had yet been able to clone human cells as Wilmut had with sheep cells and keep them alive in culture as Thomson had.

That is, until Hwang. Publishing his breakthrough in an expedited version of the journal *Science* on February 12, 2004, Hwang immediately threw both his lab and the tiny peninsula of South Korea into the scientific spotlight. With U.S. scientists still stymied by Bush's restrictive research policy, nations like South Korea saw an opportunity to stake their place in the emerging field. No one yet knew where stem cells would lead, or whether it was even possible to harness their power to treat disease, as many of the field's advocates hoped and promised. But it was obvious that despite their political baggage, the remarkable cells, like genetic engineering and gene therapy before them, represented the next advance in biological science. And for countries like South Korea, eager to take the express lane to scientific and economic prestige in biomedical research, Hwang—and his stem cells—was just the engine they needed. "Hwang's achievement has influenced quite a lot of Koreans," Dr. Jeong Bin Yim, director of the Bio-MAX facility, a newly formed government-funded state-of-the-art research institute that would serve as the hub of Korea's biomedical and stem cell research efforts, told me in 2005. "He is a Seoul National University graduate; he is a pure Korean scientist. Most Koreans know who Hwang is, and understand that he has done something very important."

At the time, Bio-MAX was still only a facility-in-the-making, a construction site outside Yim's window on the Seoul National University campus. It was only one of a long list of new ventures that the Korean government and private corporations were launching on the shoulders of Hwang's achievement. Already, the nation's leaders were building high hopes for their homegrown scientist, envisioning a bright future for their tiny country in the highly competitive and potentially lucrative field of stem cell science.

To both the Korean government and its burgeoning corporate and commercial interests, Hwang's stunning announcement gave each of them what appeared to be a can't-miss investment, both financially and scientifically. His achievement was being heralded by experts around the world, and media reports were pointing to Korea as the new hub of stem cell innovation. Leading stem cell scientists from around the world, including Wilmut, Eggan, and Daley, paid visits to Hwang's lab, their pilgrimages a combination of homage and hope for a potential collaboration with the scientist.

There was also a David and Goliath aspect to Hwang's story that was irresistible to the Korean populace. At the time, the entire country could boast only a dozen stem cell researchers, if that, compared with the hundreds in the United States. And while the U.S. government was no longer funneling money into human embryonic studies, the National Institutes of Health was still supporting stem cell work on adult stem cells to the tune of two hundred million dollars a year. Hwang's lab was making do on a million a year and still managing to trump the field. For the fiercely proud and driven Koreans, Hwang became a patriotic symbol of triumph, the humble vet who had bested the best minds and the richest facilities in the world with the most significant medical advance of the century.

★ ★ ★

During my first visit to meet him in the spring of 2005, Hwang sent one of his graduate students to accompany me on the forty-five-minute ride from the hotel to the university. In his second year in Hwang's lab, Ok Jae Koo, a quiet, friendly twenty-one-year-old, admitted that he considered himself lucky to be part of the most talked-about science team in the world. "He is a like a movie star," he told me of his mentor, still astonished over the surreality of it all. When Hwang was spotted walking around campus, awestruck students frequently stopped the professor for his autograph. Children's books retold Hwang's story—how he grew up tending cows, and performed so brilliantly in school that he gained admission to SNU—as a lesson in the virtues of hard work and perseverance. He became both a symbol and a representative of a proud nation eager to wedge its way into the scientific annals in the brand-new field of stem cell science.

As the morning traffic built up around us, Koo told me the story that Hwang was already turning into something of a signature in his numerous interviews. "The professor says we are good at manipulating the cells because we use metal chopsticks," Koo said. It was a story I'd hear again from Hwang—his breezy, sound-bite answer to the question that would, in another year, prove to be far more difficult for him to answer.

When I arrived, I was shown into Hwang's office—a small, crowded room with a single window—which also doubled as his assistant's cubicle. Crammed near the door, oddly, were a small sink and mirror, along with two sofas arranged around a low table. There was no private washroom for Korea's most celebrated scientist, apparently, but he did have his own sink for freshening up and brushing his teeth after meals, a habit about which he was apparently obsessive. His bookshelf included Christopher Reeve's *Still Me*, as well as texts on veterinary science and cell biology. Sitting in small frames on the shelves that blanketed an entire wall of the small space were pictures of Hwang with a cow—the first successfully cloned Korean species—and various plaques lauding his scientific prowess. Apologizing for his poor English, Hwang invited a collaborator, Dr. Sung Keun Kang, another veterinarian in the department, to begin answering my questions.

Kang, a bespectacled, friendly man in his thirties, seemed well versed on the ins and outs of nuclear transfer. As a coauthor of Hwang's 2004 paper, he detailed for me what he described as the innovations that the team had made in order to achieve their success with the human cells. He fleshed out some of the details from the *Science* publication, explaining how they took eggs donated by Korean women, removed their nuclei, and replaced each one with a donor's skin cell. He brought up the chopsticks—how the dexterous and steady hands of the students gently squeezing the delicate eggs was the key to keeping the eggs intact and more willing to welcome their newly donated cargo. He then described how they played with waiting for different lengths of time between inserting the nucleus and activating it with a special brew of chemicals to start dividing. They waited for two, four, and six hours, and found that two hours was ideal for pushing the egg to transform itself into a dividing embryo. Kang proudly explained how after experimenting with different cocktails of chemicals the group also honed in on the best activation protocol for the human eggs. "We achieved maximum activation of 29 percent of eggs, which is almost what we obtain from bovine and porcine oocytes," he said. Nineteen of the sixty-two oocytes that were successfully fused with their new genetic cargo began dividing.

As Kang talked, he became more excited, pointing out exactly where in the process he and his colleagues hoped to boost the efficiency even more. They had used 242 eggs to obtain one supposedly viable stem cell line, making the nuclear transfer process much less efficient than the already inefficient process of using IVF embryos.

"From IVF embryos, we can derive at least two stem cell lines from every five embryos," Kang told me, making it clear that the team was on par with the rest of the world as far as its stem cell derivation skills. Their success, he claimed, lay in how they extracted the stem cells from the ball-shaped blastocyst. Some researchers were using immunosurgery—targeting the outer layer of blastocyst cells, which did not contain any stem cells, with antibodies that would recognize them, bind with them, and then kill them off, leaving behind the stem cells. Others relied on a waiting game, holding off on doing anything until the blastocyst naturally "hatched," or attached itself to the uterus, at which point researchers mechanically sliced off the inner cell mass containing the sought-after stem cells. Hwang's group used an amalgam of both of these methods, allowing the age and condition of the blastocysts to dictate which strategy would yield the most cells with the least amount of disruption.

Kang then went on to describe the most significant step in the process: confirming the identity of the stem cells. Like Thomson and Gearhart, in his paper Hwang detailed how the cells he grew in culture were able to form teratomas containing the three primitive germ cells that eventually give rise to all of the body's tissues. On paper, at least, it looked as if he had generated stem cells.

★ ★ ★

Following the biology lesson, Hwang was ready and eager to lead me through a tour of his lab. What I saw during that visit was the animal half of his operation, which was confined to one side of a hallway that bisected the sixth-floor lab space. On the other side was a far more restricted area visible only through a small window cut into the hallway wall. It wouldn't be until my second visit, several months later, that he would allow me into this sanctum of human embryonic cells.

On that first visit, the animal lab was buzzing with activity; in the first room, about fifteen identically clad technicians in blue gowns were busy with cow and pig ovaries. Hwang received fresh ovaries from slaughterhouses three times a day. The organs arrived in thermos-like containers and were first cultured for twelve to forty-four hours to help the oocytes inside to mature. Then the students extracted the eggs, by hand, in a laborious and tedious process.

In a second, darker room illuminated only by the glow from eight heavy microscopes, a handful of students were putting their chopstick skills to good use, gently pricking the animal eggs and then squeezing out the genetic material. The eggs seemed to lose their filling the same way a jelly doughnut does with a bite. Using the same slit from which the gene-laden nucleus had just exited, Hwang's students then injected a new cell, nucleus and all, to replace the oocyte's missing DNA. The combination was then zapped with a jolt of electricity to fuse the disparate parts together. Chemicals then nurtured the hybrid to begin dividing. In five to six days, presto! A dividing embryo that was ready to be implanted into a surrogate animal's womb in hopes of creating a clone.

Because nurturing cells is a twenty-four-hour, 365-day-a-year process, "there is no holiday in our lab, even on Korean Thanksgiving or New Year's Day," Hwang said. Then he laughed with the easy humor that endeared him to all who had met him. "A foreign scientist once called my lab the troops, and me the commander."

It's an interesting metaphor, and one that would prove prophetic in the following year, as revelations about misconduct in the lab exposed the extent to which the hierarchical structure of academic research departments and the cultlike devotion that Hwang inspired among his lab members contributed to a sense of invincibility that may have led to his downfall. For most of the students Hwang was like a second father, often referring to them with the Korean word for "children" and worrying about the extended hours they spent cooped up in the sixth-floor labs.

Jiho Choi, a master's candidate at the time, remembers Hwang's generosity when we talk about his former mentor years after Choi, now earning his Ph.D. at Harvard, graduated. "He was more than just a principal investigator," he says. "He paid the tuitions of students and also a small stipend for room or a dormitory to live in for people who came from [distant] provinces. He did everything for us." Hwang's goal, says Choi, was to provide an environment where students would feel encouraged and inspired to focus on research without getting distracted by practical concerns, so he could seed a new generation of stem cell scientists in Korea. "If I had some problem or a difficult situation, then I could talk to him," Choi says, still using the Korean honorific *sunsengnim* to refer to Hwang, a sign of respect for an accomplished and well-liked superior.

★ ★ ★

Over lunch at the faculty's Chinese restaurant on campus, Hwang explained why, despite the added attention from a proud nation, he preferred to keep his lab small—his core team, not counting collaborators from other departments in the university, remained about the same size as it had always been: forty-five members. "In our opinion, this work must be done by hand," he said. "This kind of work, you need to insert the spirit, and use not only machines, but the heart and spirit. You need a human touch."

It was the first hint of spirituality Hwang had expressed. Like many Koreans he was a devout Buddhist who often traveled to temples for ritual spiritual cleansings and moral guidance. And as is the case with many scientists in the stem cell field, he had made his own peace with the apparent conflict between what he did in the lab and his spiritual perspective. In his view, he explained, the purpose of his work—to improve agricultural stocks, in the case of his animal studies; and to contribute to new treatments for disease, in the case of the human stem cell work—made the moral challenges justifiable.

As we picked our way through the traditional Korean meal of rice, *banchan* side dishes, and soup, I asked him about the one issue that puzzled many of the Western scientists struggling to do the same thing he was. Where, I asked, was he getting so many eggs?

Hwang talked about the unbelievable amount of support and attention his lab received in the aftermath of the breakthrough. Even before the paper was published, he told me, the generosity of women eager to contribute to the research provided him with enough eggs to practice and perfect the finicky technique of nuclear transfer with human cells. In complete contrast to the challenges the Harvard team was facing, Hwang, it seemed, had more than enough human volunteers for his stem cell studies.

★ ★ ★

Following lunch, Hwang had planned a trip to the pig farm, where he intended to perform one of the hundreds of embryo transfers that his team conducted every day to grow clones. "He can do the transfer in twenty seconds," Byeong Chun Lee, a close collaborator, informed me, clearly impressed. "Watch him. You will be surprised."

But going anywhere with Hwang was not a simple thing. Like the president, Hwang moved with a security detail, including two bodyguards assigned to him twenty-four hours a day. Aside from driving him to the lab every morning, they also accompanied him on trips overseas, where he was becoming a popular conference speaker. During the day, they haunted Building 85, alternating between the elevator on the sixth floor and the garage beneath the building, where Hwang's car was always at the ready. They came courtesy of none other than former president Roh Moo-hyun himself. A little embarrassed, Hwang admitted that he initially refused them, but apparently it's difficult to turn down the president.

As soon as we emerged from the cramped streets of the city, the landscape began to change. The urban collage of neon, glistening steel, glass, and concrete was gradually subsumed by thicker growths of trees and what seemed like endless stretches of farmland. Lush rice paddies blanketed much of the scene that rushed past our windows. As the last vestiges of Seoul slipped by, the increasingly pastoral landscape must have reminded Hwang of his own childhood home in the countryside, and he started to open up about his remarkable road to science superstardom.

Hwang was born in 1953 in Chungnam, South Korea, a rural province three hours outside the country's capital. It was just after the Korean War, and like many families not fortunate enough to be living in metropolitan areas such as Seoul or Busan to the south, where jobs and economic recovery came more quickly, the Hwangs barely had enough to eat. His father passed away when he was five, leaving his mother to raise her brood of six children alone.

For families such as theirs, farm animals were the key to survival. Young Woo Suk quickly learned to love his duties caring for the three cows the family managed to obtain. "Even those," he admits, "were not really our cows; we rented them from our rich neighbors." The young boy, just tall enough to look a cow in the eye, felt a special affinity for the animals, and would talk to them constantly, vowing always to look after them. There was something about their serenity that appealed to him, and the daily task of feeding and cleaning up after the animals may have given the young Hwang a sense of rootedness. It was during those afternoons in the company of cows that Hwang began to dream of becoming a veterinarian.

In high school, Hwang's grades were good enough for his teachers to suggest that he consider medical school at Seoul National University, the nation's most esteemed college. Competition for admission to SNU was fierce, but the studious boy earned one of the coveted spots in the freshman class of 1977. In one of his first acts of stubborn defiance, however, Hwang refused to become a doctor. Instead, he enrolled in SNU's veterinary school, still determined to study animals. After earning his Ph.D. in animal reproduction, he left Korea for the first time, hopping over the Sea of Japan for a fellowship at Hokkaido University. As a fellow, Hwang became interested in the delicate process of embryo splitting. It was something nature did on its own every once in a while, in producing identical twins. But Hwang wanted to bring more consistency to the process. It was his introduction to cloning.

When Hwang returned to Korea, he came back to his alma mater. Growing up in Chungnam, he says, there were only two things he had desired—to become a scientist, and to become a professor at SNU. He arrived with valuable skills and an understanding of how to optimize the reproductive process. His goal was to improve the yield of milk and meat production in cattle—something farmers and agricultural experts had been seeking for centuries. Hwang thought he could do it with the help of technology, by applying what he had learned in the lab to the barn. He wanted to clone the most desirable cows—the "elite" animals—and thereby seed the Korean cattle population with only the premium producers of milk and meat. In 1993, he produced the first test tube cow, using in vitro fertilization to inseminate a bovine egg in a dish and then transferring the resulting embryo into the womb of a surrogate cow for gestation.

Two years later, Hwang managed to successfully split an embryo and produce the first cloned cow in Korea. But embryo splitting wasn't truly cloning. Technically speaking, it was twinning with a hand from science. Then came news from Scotland of Wilmut's success in producing Dolly, a true mammalian clone who was the exact genetic copy of an adult ewe. As it did for so many scientists in the field, Dolly's existence had a profound impact on Hwang's thinking at the time. Her birth convinced him that more wide-scale cloning of animals, using nuclear transfer, was possible.

Applying the technique to cow eggs, in just two years Hwang had his own cloned cow via nuclear transfer. But he never published his experiment in a scientific journal, choosing instead to announce the birth in the press—the first cloned Korean cow, Hwang said, as a gift to the Korean people—complete with photo ops. Other scientists could neither confirm nor deny nor learn from his attempts.

Halfway into our journey back to the labs, I asked Hwang about the challenges of keeping human embryonic stem cells happy—of maintaining the cultures so they remain pure and stemlike, resisting the urge to differentiate into some other cell.

He nodded, reflected a bit, and then responded. "I always worry about contamination," he said. "Especially in the human stem cell lab, because I had a painful experience. Some human embryonic stem cells became contaminated, and were discarded. It took six months to get over the contamination."

It was a surprising revelation, and if true, a potentially telling one given the events that transpired over the next six months. Losing stem cell lines is a big deal, but losing human embryonic stem cell lines is a disaster. Human eggs are difficult enough to come by for a research lab, and even Hwang's efficiency at the time in turning those eggs into stem cells was pretty dismal: He reported in 2004 that it had taken 242 eggs to produce one stem cell line.

I asked how many lines were destroyed.

Some, but not all, he answered. Construction on the ventilation system in the building had apparently contaminated the relatively sterile lab; pathogens in the air that wouldn't bother the average person could settle onto a cell culture and wipe out the colony.

Following that mishap, the government stepped in, quickly renovating the space with a more reliable ventilation system and creating the separate, strictly controlled space that became home to the human side of the stem cell lab. At that point, anything Hwang wanted, Hwang got.

It was astounding that a scientist could hold that much influence with the country's top political leaders. But that was the nature of

Hwang's new stature in Korea. He was by then no longer simply a scientist, or even a professor at the nation's most competitive university. He was a national hero, and for better or worse, an icon both for what Korea was and for what it wanted to be. A month after my visit, the title of "supreme scientist" of Korea was bestowed upon Hwang, a new honorific concocted by the government to recognize his efforts in leading the nation onto the world's scientific stage. In Hwang, President Roh saw a chance for Korea to finally step up to the head of the line and bring the rest of the world along—behind it. With stem cells, Korea had an opportunity to establish an infrastructure and expertise that Roh and his administration thought would yield dividends both financially and politically in the global economy.

But it wasn't just political opportunity that Roh seemed to see in the veterinary scientist. Roh, like Hwang, was a self-made man who had grown up in poverty, and worked his way through law school. In Hwang he had found a brother, a kindred spirit who had earned his way into the upper echelons of power and influence in a hierarchical society where such feats were rare, if not unheard of.

"President Roh said to me, 'I want to be recorded as the president who concentrated on developing science and technology,'" Hwang told me, before adding with a trace of boastful pride, "I promised him [he would] be."

At the time, I didn't appreciate the full import of Hwang's words. But they would prove pregnant with deeper meaning. In 2009, Roh committed suicide in the midst of a bribery scandal. And that promise, and the expectations inherent in the pledge, may have been the beginning of the end for Hwang.

★ ★ ★

"I got the impression he was under tremendous pressure," says Insoo Hyun, who spent the summer of 2005, the height of Hwang's popularity in Korea, in the cloning lab. A professor at Case Western Reserve University, Hyun traveled to Seoul on a Fulbright-Hays Faculty Research fellowship to study the informed consent process in Korea. Like many others, he was interested in learning how the Koreans were able to obtain so many eggs from female volunteers for their stem cell work, and how the informed consent process was set up to protect donors. Why, he wondered, had the breakthrough occurred in Korea? Why were so many women there willing to donate eggs for research?

Easygoing and approachable, Hyun, a Korean American who understands Korean and admits to speaking it only haltingly, was immediately embraced by Hwang as a countryman. "His first question to me was, 'Do you speak Korean?'" he says. "I said, 'I don't speak it very well, but I understand most of it.' He said, 'Good!' and after that everything was all in Korean."

Even during his short time in Seoul, Hyun witnessed the dizzying heights to which Hwang's popularity had climbed, and the burden that came along with that fame. A simple dinner at a restaurant evolved into a security operation worthy of the military, as Hwang was escorted out the kitchen door while Hyun and the rest of the party distracted photographers at the front door.

Unlike the way scientists are funded in the United States—project by project, with each new experiment or hypothesis requiring a new grant and a new proposal—in Korea, the government funds the scientists themselves. That is, it supports a particular researcher's lab and keeps him afloat to test whatever theories he chooses to challenge, or whatever experiments he wants to perform. "What they do is say that we will put everything in the Hwang basket," says Hyun. "We fund him—just him and his lab, and put a lot of money into him. And the funding for Hwang came from the technological and business development parts of the government—not Health and Human Services. So clearly there was a business or payoff angle to the research."

Hwang was certainly reminded of this every day as he walked through the bright blue door leading to his office. The sign above his name on the wall read POSCO CHAIR, acknowledging the endowed chair he held in the school of veterinary medicine—the first such title of any

kind at Seoul National University. Executives at POSCO, Korea's leading construction company, had wanted their own piece of the Hwang
aura and had agreed to support Hwang's salary to ensure the scientist
would stay in Korea and not be lured overseas to pursue his stem cell
studies.

Before my first visit ended, Hwang asked how long I would remain
in Seoul. He didn't reveal why, but he invited me to a press conference he
was planning the following week at, of all places, Incheon International
Airport, an hour outside of Seoul. He was charmingly coy about the
details, saying only that I would be "very interested," and "surprised."

★ ★ ★

"He just landed," I heard the broadcaster announce on his cell phone to
his producer on the other end. "He's on his way."

It was 4:20 P.M. on May 20, and Hwang was already twenty minutes
late for his own press conference. The television reporter on his cell
phone was clearly excited, pacing between the small Magnolia Room in
the airport and the hallway outside. Anytime Hwang appeared, it was
news. Cameramen lined the walls of the room, and about two dozen
journalists, many of whom were permanently assigned the "Hwang
beat" to cover the scientist's every move, weren't surprised he was late.
They seemed accustomed to the delay.

Half an hour later, Hwang finally made his way down the hall and,
in true rock star fashion, was ushered in by his ever-present bodyguards
and a knot of photographers pushing and shoving for the best shot.

As he entered the room, Hwang greeted most of the reporters personally, shaking their hands and touching their arms in recognition. It
was a very Western way of acknowledging people—a departure from
the more reserved bow and physical distance Koreans normally use in
professional settings.

Hwang took his seat in the front of the room, and addressed the
room in Korean, speaking slowly, even poetically at times, of the work
that would, in just a matter of months, prove to be nothing but fiction.

The week before the press conference, the journal *Science* had published the much anticipated sequel to Hwang's 2004 feat. This time, he had used the cloning process first to clone, then to extract embryonic stem cells from patients suffering from spinal cord injury and type 1 diabetes. Even more astounding, he said, was that in doing so, he had boosted the efficiency of the finicky process tenfold, creating eleven stem cell lines from 185 eggs.

Up to that point, it wasn't clear that cells from patients could be cloned in the same way healthy cells could. Theoretically, there was no reason why they couldn't, but practically, nobody knew whether diseased cells could be reprogrammed and cultured at all. With his second paper, Hwang had proved that they could.

In addressing the small group, Hwang looked squarely into the cameras. If he knew at that point that the cells were not what he claimed them to be, he showed no signs of doubt.

"We asked ourselves, 'If we take cells from a patient, can we make stem cells?' It has never been done before," he began. "Even human stem cells are now grown on animal feeder cells, which means they cannot be used in humans because of contamination with animal proteins. So I used cells from the patient himself to create a feeder. And we found that human feeder cells work better than the mouse cells to grow human stem cells. We used a hundred and eighty-five cells to generate eleven stem cell lines. This is a tenfold improvement of efficiency from our study last year. So this means that from one egg-harvesting cycle, we can generate one stem cell line. I have now shown that you can take cells from a patient, grow stem cells, and then place them back in the patient."

It was a masterful speech, saturated with Hwang's hallmark mix of deference, charm, and respectful confidence.

Six months later, he would face many of the same reporters again. But this time, the crowd would not be as friendly. The cells he described in May had by then been exposed as a fraud, and the next time he came before the cameras, Hwang would be making a very different kind of speech. He would apologize, to the nation and to the world, for his part in a stunning case of scientific fraud. Hwang's integrity as a scientist and his entire stem cell operation would then collapse around him.

★ ★ ★

Sometimes it's the most obvious question that never gets asked, and the most basic assumptions that prove to be the most dangerous. It's true in any discipline, but most of all in science, where every query is an opportunity and every investigation a window that can spill light into the darkness of the unknown.

Hwang's astounding success in generating the first human embryonic stem cell line from a cloned embryo depended heavily on one thing—eggs. Or more specifically, a ready population of women willing to donate eggs for research. For his 2004 breakthrough, Hwang used 242 eggs to generate one stem cell line. Describing the process in his *Science* paper, he noted that these eggs had been harvested from sixteen women who had signed informed consent forms, were aware they were donating their eggs for research purposes, and were not compensated for their donation.

In May 2004, just three months after *Science* published the Hwang team's milestone results, questions about exactly how he had been able to convince so many women to donate eggs started to surface. Curious, David Cyranoski, a journalist with *Nature*, a competitor of *Science*, began asking some of Hwang's coauthors and members of the hospital review board that had approved the study, at Hanyang University Hospital, about where the eggs had come from. To his surprise Ja Min Koo, a coauthor on the paper admitted, in a burst of patriotic pride, that she and a colleague had both donated eggs for the experiment.

The following day, however, Koo called Cyranoski back to change her story. Blaming her poor English, she told him she had not donated any eggs at all and what she had meant to say was that she would be more than happy to donate if she could. But institutional review boards, the hospital committees that oversee human research, prohibit subordinates from participating in research conducted by their lab leaders, to avoid potential coercion. Koo would have been doubly ineligible to donate, since she was not only a member of Hwang's staff but also a coauthor on the paper, and would therefore be in a position to gain, both financially and professionally, from the study.

Cyranoski wasn't convinced by Koo's second story. In the initial interview, she had described in detail the hospital where she went to donate the eggs, and discussed her desire, as a Korean woman, to help Hwang because she already had two children. He next attempted to confirm the source of Hwang's 242 eggs, by contacting the Hanyang IRB and Hwang himself. Hwang, according to Cyranoski, denied that Koo had donated eggs to the study, but did not provide evidence to support his denial.

On May 6, *Nature* published Cyranoski's story raising questions about whether the eggs in Hwang's study had been procured ethically. In it, he cited Koo. The allegations prompted the Korean Bioethics Association and the People's Solidarity for Participatory Democracy to call for an investigation into the ethics of Hwang's 2004 paper.

The following week, Hwang continued to deny any wrongdoing, bolstered now by a letter of support from the head of the Hanyang IRB, Moon-Il Park, a professor of obstetrics and gynecology. Hwang noted in an e-mail to Korean reporters that after he was questioned by Park, the IRB member said he was confident that no one from Hwang's group had donated eggs.

Still, suspicions have a habit of growing, and Hwang's once-spectacular success story was now showing signs of weakness. Ethicists, many of whom, because of religious reasons, were never comfortable with the idea of research on human embryos, were beginning to voice concerns over how Hwang had carried out his study. The questions about the egg donation now provided them with an outlet for expressing their reservations—and signified a potential crack in Hwang's superscientist armor. Young-Mo Koo, a professor of medical ethics at the University of Ulsan College of Medicine, told *Nature* that he felt "deeply betrayed" by his colleagues' research.

That was enough to push the Korean Bioethics Association into action. In a statement issued on May 22, the group urged Hwang and the IRB to clear up lingering questions over how egg donors were recruited and whether they had been paid. The association also sent a letter to *Science* editors, declaring that Hwang's team may have proceeded with the cloning study without proper vetting of the ethical and moral questions it raised.

Hwang immediately responded by dismissing the allegations of misconduct. "They are trying to denounce the research," he said.

Hwang appeared confident that his experiment was both just and justified. In anticipation of such scrutiny and controversy in working with human embryos, in late 2004, after the furor over his initial paper, Hwang had reached out to Kyu Won Jung, a lawyer and ethicist at Hanyang University. Jung had published a paper in 2001 supporting stem cell research, an unpopular position with the majority of bioethicists in the country, who had adopted a more conservative position against the studies on embryos.

Hwang told Jung he wanted an ethicist to help write an impeccable informed consent that would stand up to scrutiny for his second experiment with patients' cells. Jung agreed, although he was a bit wary of the ease with which Hwang seemed to think it was possible to make human embryonic stem cells. In a debate with Hwang over the subject, Jung recalls that most experts predicted it would take a decade or more before scientists would be able to grow human embryonic stem cells using nuclear transfer. "But Hwang said it would take less than three years," he says. "I thought it was too soon."

Jung insisted on a rigorous informed consent process that required three separate interviews to ensure that women were donating for the right reasons, had not been coerced, and did not feel obligated to give for financial, professional, or misguided patriotic reasons. In Jung's process, which he described in a paper published with Hyun in the *American Journal of Bioethics*, women would have to be interviewed by a member of Hwang's scientific team, Jung (or another bioethicist), *and* an obstetrician-gynecologist.

Hwang and his team, Jung learned later, weren't always faithful to the stringent procedure that the ethicist had laid out. Jung himself interviewed several women, and ended up rejecting some when he felt they were giving for the wrong reasons. But Hwang's collaborators at Miz-Medi Hospital, where the women were coming to volunteer, were not as vigilant. "The problem is they did not follow the informed consent," Jung says.

Hwang, meanwhile, was enjoying the peak of his national popularity. Awards and invitations to speak at events, scientific and otherwise, poured in; he had become both celebrity and dignitary with several deft squeezes of the metal chopsticks. Korean Air offered both him and his wife a lifetime of complimentary travel anywhere the national carrier flew, turning the airline into his virtual private jet. During our meeting, he proudly showed off the latest cell phone from Samsung, not yet on the market, that the heir to the electronics giant had personally presented to Hwang as a token of appreciation for his achievements.

But it turns out Hwang may have been exploiting this popularity to deflect attention away from the nagging questions about his egg donors. In June, he announced that South Korea would be home to a new stem cell bank, the first of its kind, that would grow and house stem cell lines for anyone in the world interested in using them to treat disease. He envisioned the bank as a global resource for linking patients in need of replacement cells with their closest possible stem cell match.

Even as the questions about the egg donors continued to swirl around Hwang, the day the World Stem Cell Hub began accepting applications both online and in person from patients willing to donate their skin cells in October 2005, it received so many requests that the website crashed. By the end of that first day, thirty-five hundred patients had registered to become donors, all from Korea. Many, some in wheelchairs, had made the trip to register in person at Seoul National University Hospital, where the Hub is based, eager to play their part in what they hoped would be a history-making effort to cure their diseases.

★ ★ ★

During my first visit, in 2005, even before his second breakthrough paper on patient-specific stem cells, Hwang was winnowing down the list of scientists who were clamoring to collaborate with him. He provided me with a neat list of his top picks, complete with e-mail addresses and affiliations: Ian Wilmut; Lorenz Studer, a neurologist at Memorial Sloan-Kettering Cancer Center, in New York; Roger Pedersen, now at Cambridge University, in England; and Curt Civin, a cancer researcher then at Johns Hopkins University, in Baltimore.

At the top of Hwang's list was Dr. Gerald Schatten, a biologist at the University of Pittsburgh. For nearly two years, Hwang and Schatten had been scientific "brothers," developing a close professional and personal friendship. It was a fast and furious courtship that earned Schatten a coveted coauthor credit on Hwang's second stem cell paper, despite only acting as a consultant. Schatten was also chair of the international board of directors of the World Stem Cell Hub.

The honeymoon ended in a contentious divorce, however, when Schatten tersely ended his collaboration with Hwang in a cryptic press release on November 12, 2005, just a month after the launch of the World Stem Cell Hub. Declining to explain his reasons, Schatten stated only that "information came to my attention suggesting misrepresentations might have occurred."

His words were like a live grenade dropping onto the stem cell community; everyone panicked. Researchers who had committed to participating in the World Stem Cell Hub put their plans on hold, and scientists hoping to obtain stem cell lines from Hwang for study also had their hopes dashed as few were willing to move forward until the cloud over Hwang had cleared.

No one seemed more stunned than Hwang himself. Fresh off a plane, he was greeted upon arriving in Seoul by the familiar gaggle of reporters and cameras in the terminal. One asked about Schatten's sudden change of heart; it was the first time that Hwang and his associates, including his deputy, Ahn Curie, had heard the news.

Anxious, they immediately called their new friend Hyun, who had since returned to the States. His phone rang at five o'clock on a weekend morning in his Cleveland home. "They asked me, 'What do we do,

what do we do?' I told them, 'I can't advise you because I don't know
what happened. But whatever happened, if you did something wrong,
apologize right away and set the record straight and I'll talk to you when
things calm down.' Well, they never calmed down, and they never contacted me again."

Instead, Hwang and his close-knit coterie of advisers, which included
Curie, Kang, and Lee, opted for a different approach: Maintain their
innocence and deny, deny, deny.

"I know nothing except that Schatten had announced a separation,"
Hwang told reporters in Seoul when he finally addressed the public two
days after the U.S. scientist's stunning announcement. Schatten maintained his silence (and still has not discussed with reporters his reason
for ending the relationship with Hwang), further adding to the intrigue.
Hwang promised he would tell his side of the story once he'd had a
chance to look into Schatten's concerns and straighten things out. "I will
tell everything when the right time comes," he said.

★ ★ ★

Just after midnight on June 1, an e-mail arrived in the tipster account at
a local Korean television station. Producers of an investigative news show,
PD Notebook, had set up the account to field information, often anonymous, on juicy scandals of any kind, from ethical breaches to bribery to
simple acts of bad behavior, which they would then investigate and film
for compelling exposés that aired in prime time. Around the time the
press was fawning over the announcement of the new World Stem Cell
Hub, a researcher, apparently from Hwang's lab, e-mailed the account,
according to an article published in *Science* months later. Bo Seul Kim,
one of the producers, opened the message. The author, identifying himself as a member of Hwang's team, admitted that he was increasingly
uncomfortable with the attention that Hwang was receiving, given what
he knew of the work that had led up to the landmark 2004 paper. Then
the tipster dropped a bombshell: Hwang was lying about the eggs. Some

of the eggs used in the study had indeed come from women in the lab, and he had an e-mail from one of them confirming that Hwang had asked her to provide them.

Through the whistleblower, another producer on *PD Notebook*, Hak Soo Han, worked to validate the claims. Han was able to convince other members of Hwang's team to expose the inner workings of the lab. He also brought in three scientists familiar with cell biology to begin combing through Hwang's scientific papers. These consultants might have been the first to scrutinize Hwang's claims carefully enough to notice significant anomalies. One of the first to stand out was Hwang's failure, in the 2005 paper, to use the teratoma test to document that each of the eleven stem cell lines he claimed to have generated from patients' cells were indeed bona fide stem cells. In the paper, Hwang's group describes forming teratomas from only two stem cell lines.

But it still wasn't clear whether that simply represented a conscious decision to streamline the data presented in the paper or whether it was a hint of scientific obfuscation. Taking a stab in the dark, one of the advisers wondered whether the stem cell lines had simply been generated from IVF embryos, and not through the difficult process of somatic cell nuclear transfer, as Hwang had claimed. To prove that, however, the conspiracy theorists needed to get their hands on both the stem cell lines and the DNA of the donor from which they had been derived.

They put in a request for DNA data from fifteen stem cell lines created at MizMedi Hospital, which had provided Hwang with the eggs from some of the IVF embryos they maintained once couples had donated them for research. Through some subterfuge, the scientists managed to convince one of the members from Hwang's team who was assisting the investigation to sneak out a sample of one of the stem cell lines described in the 2005 paper. When they sent a sample of those cells to a lab for testing, they learned that its DNA matched that of one of the fifteen lines from MizMedi. That could mean either that Hwang had not created any stem cell lines using nuclear transfer, or that he had not created as many lines as he claimed.

Han was not in Seoul when he received this stunning news by e-mail on October 19. He had flown to Pittsburgh to interview Sun Jong Kim, one of Hwang's researchers who had been "loaned" to Schatten's lab so that Schatten could learn the cloning technique that the Koreans had perfected. Han's timing was perfect. Armed with the test results, he resorted to one of the oldest tricks used by policemen the world over to turn a snitch. Even with the report of the DNA match, Han still couldn't be sure that Hwang had actually misrepresented his work. But he told Kim otherwise, convincing him that he already had enough evidence to have Hwang arrested for fraud. Han also used a hidden camera to get Kim's response on film.

The tactic worked. Kim sang. He admitted that Hwang, now revered by the entire nation as a symbol of pride, integrity, and achievement, had instructed him to doctor photographs submitted to *Science* to make two stem cell lines appear as eleven separate ones.

Kim has never spoken publicly about his involvement with Han or about the circumstances of his confession. But according to members of Hwang's lab, it was likely the basis of Schatten's abrupt decision to distance himself from Hwang.

★ ★ ★

As the questions continued to build, Hwang made a brief escape. A deeply religious man who had considered himself Catholic before converting to Buddhism, Hwang, as so many Koreans do in times of crisis, retreated to a Buddhist temple.

Only Hwang knows the true purpose of his journey, and whether it was meant to restore or repent. He emerged, apparently renewed, with a handmade bracelet. A monk, he said, had given it to him as a reminder of his spiritual strength. "He told me that I should keep in mind the mercy and the true love of Buddha," Hwang said. "He gave me the insight and wisdom that I need to overcome challenges and crises in life."

Whether or not Hwang had any inkling of the challenges to come still isn't clear. A week after Schatten's break, Hwang's chief collaborator at MizMedi, Sung Il Roh, admitted that some of the egg donors whose oocytes had made history in Hwang's 2004 study had been paid for their services. All together, said Roh, he shelled out 1.5 million won, or fifteen hundred dollars, of his own money, to each of twenty women who agreed to provide eggs. "I did this work without consulting Dr. Hwang because I felt it would help in the search for new ways in finding cures to untreatable diseases," Roh said.

That same evening, at eleven P.M., Munhwa Broadcasting Corporation (MBC) aired the result of its collaboration with the tipster in an explosive investigative documentary on the egg donation in Hwang's research. The producers interviewed women who claimed that they had sold their eggs to MizMedi, the hospital where Roh worked. Many of those eggs were then supplied to Hwang for his stem cell studies. Some of the women interviewed, whose names were withheld and whose faces were disguised, said they hadn't known that their eggs would be used for stem cell research—contradicting Hwang's assertion that the eggs had been obtained from donors who had signed IRB-approved informed consent forms and were fully aware of what they were doing.

Such apparent disregard for the ethics underpinning each scientific decision raised all sorts of red flags, and only bolstered the position of the slippery-slopers. Allowing cloning in order to obtain stem cells would only provide more opportunities to exploit and coerce people and their bodies, they warned.

Within weeks, on Thanksgiving Day in the United States, Hwang publicly admitted to lying about the fact that two of his researchers had donated eggs to the study. He maintained, however, that he had learned of the donations only after the fact, and that when the women initially expressed their desire to help the study by providing their own eggs, he had refused. But the junior members of the lab had gone to MizMedi on their own, he said, allegedly using different names to become donors. In a tearful apology, Hwang acknowledged "being too focused on scientific development," and that he "may not have seen all the ethical issues related to [his] research." In an act of contrition, he resigned as director of the World Stem Cell Hub.

Stunned by the news, Jung, who had worked so carefully to craft the informed consent process, reviewed his own records of the women he had interviewed, as well as those whom Hwang's team had spoken to, who had also donated eggs. In the first six months of 2005, Jung had talked to about two dozen potential donors. Of these, he had approved only eleven or twelve. When Jung looked more carefully at the records Hwang's lab had kept on the donors for the 2005 study, he was surprised—and disappointed—to find that the strict process he had put in place had not always been followed.

The admission was Hwang's first acknowledgment of wrongdoing and all the leverage that some Korean scientists, frustrated over competing with Hwang for funding, needed to air their grievances. "There seemed to be a deep-seated resentment among other scientists that was squashed, kept back and repressed during the time of Hwang's celebratory celebrity," says Hyun. "But they are the ones that now were looking online, looking through the photos from his papers and exposing all the problems. They were combing through all of his stuff. That suggests to me that, given the funding structure in Korea of backing scientists [and not science projects], several people may have been professionally jealous, and wanted to see him fail."

They didn't have to wait long. While the government rushed to save its investment in the Hub by announcing that it would continue to fund Hwang's lab with three million dollars a year, Ahn Curie flew to the United States and Japan for a ten-day damage-control trip. When she returned, she didn't have good news.

"We cannot be optimistic at all about collaborating with overseas laboratories," Sung Myong-hoon, an official at the Hub, told reporters at a press conference. In early December, Doug Melton and Kevin Eggan, who had asked Hwang to be the keynote speaker at an upcoming conference, made the difficult decision to retract the invitation. Eggan e-mailed Hwang, diplomatically citing the investigations that must be occupying most of his time, and politely suggesting a rescheduling of the talk. "He understood," says Eggan.

Oddly, however, while the scientific community was backing away from Hwang, the Korean public only redoubled their support. Most saw him as a victim rather than a perpetrator, the target of jealous collaborators and manipulative colleagues. They believed Hwang's explanation that he kept his lab members' donations under wraps out of respect for their privacy. "In the end, I could not ignore the strong request by the researchers to protect their privacy," he stated at the press conference.

Hwang's supporters weren't content to just talk up the selfless and admirable qualities of their hero. Taking advantage of the country's highly wired social network, they took to online message boards with violent threats against producers at MBC, which had aired the original segment questioning the integrity of Hwang's egg procurement process. When they went to the streets, staging a demonstration outside the MBC building in Seoul, eleven of the twelve sponsors of *PD Notebook* terminated their funding. The show was forced to shut down for several months.

In the scientific community, however, things were different. There Hwang had gone from golden boy to outcast; his decision to lie about his knowledge of the egg donations made him a pariah. If he had lied about that, researchers wondered, what else could he be hiding? Stem cell experts were particularly sensitive to any breaches in protocol that could be perceived by an already wary public as all the evidence they needed to justify the dangers of stem cell science. These scientists knew that every misstep could be potentially fatal for the field.

★ ★ ★

In Washington, D.C., the editors of *Science*, which had published both of Hwang's milestone papers, were scrambling to determine if their review process had let important defects in the Korean papers slip through. Outwardly, the journal's editor in chief defended the publication's vetting process. "None of these allegations [by MBC] have been credible,"

Donald Kennedy said. "Until a specific, scientifically based claim against Dr. Hwang's findings is reported to us, we will not offer speculations."

That came on December 4, 2005. At 11:29 P.M. *Science* editors in D.C. received an e-mail from Hwang, alerting the journal to mistakenly duplicated images in the supporting material supplied with the 2005 paper. "We made some unintentional errors by using about four pictures redundantly," he wrote. Re-creating the paper trail of submissions and e-mail correspondence, *Science* editors determined that the error appeared in the final online version of the study but was not part of the initial submission from the Korean team. Prior to publication, the editors had asked Schatten, one of the authors, to provide high-resolution images to run with the paper. He did, after obtaining them from Hwang, he told the editors. Nobody noticed that the images were duplicates and did not match the picture initially submitted with the paper.

The next day, the University of Pittsburgh announced that it was conducting its own investigation of the paper, based on the mix-up with the images.

Almost immediately, Hwang entered the hospital at Seoul National University Medical Center, for exhaustion and stress. Pictures of the scientist uncharacteristically unshaven and unkempt, lying with his eyes closed on a hospital bed, flooded the Korean media outlets.

As the fifty-two-year-old remained sequestered in his sickroom, Schatten wrote a fifty-six-page letter to *Science*, detailing transgressions Hwang had committed that had come to his attention. Citing the violations, Schatten requested that *Science* remove his name from the 2005 paper, but the journal's policy holds that no single author can retract his name on a paper after attesting to the validity of the contents at the time of submission.

Contacted by *Science*, Hwang and his team admitted to minor clerical errors in the 2005 report, as well as a major correction concerning the egg donors. Following Roh's admission that he had provided eggs for the study that had been paid for, *Science* printed a clarification noting that some donors had received about fourteen hundred dollars for their eggs.

But the biggest shock was yet to come. On December 15, Hwang's controversial partner and egg procurer, Roh, visited the embattled vet in the hospital. Roh emerged with the jaw-dropping news that Hwang had admitted to faking the stem cells. "I heard some things that I haven't been aware of . . . that there are no embryonic stem cells," said Roh.

Hwang took to the mike the following day, still maintaining his innocence.

"Our research team made patient-specific embryonic stem cells, and we have the technology to produce them," he said, defending the work. He promised to thaw the remaining five lines and prove that they matched the skin cell donors from which they were made.

Unfortunately, Hwang was wrong. An investigation by Seoul National University officials found that the remaining cell lines matched IVF embryos, not the skin cell donors in Hwang's 2005 study. Moreover, the 2004 cell lines could not be confirmed as originating from nuclear transfer either.

It was a shocking admission. The breakthrough on which an entire field depended was now apparently a sham. The stem cells that Hwang claimed to have nurtured from human cells, for the first time, had never really existed.

In the waning days of 2005, Hwang himself finally admitted the truth. "I feel so crushed and humiliated that I hardly have the energy to say I am sorry," he said through tears during a news conference. He maintained, however, that the fault was not entirely his, but originated with his collaborators at MizMedi, who had deceived him into thinking they had successfully coaxed stem cell lines to grow from the embryos that Hwang's team had created. "I one hundred percent trusted what they told me. Now I believe that they completely cheated me."

Throughout the investigations and court cases that followed, which would stretch out for another three years as Hwang was brought to trial for falsifying his results and embezzling government funds, he insisted that his team had the skills to clone human cells in order to generate stem cell lines. To the end, he never wavered from this belief. The truth, however, is that while they may have had the skills to clone the cells, they lacked the knowledge and the experience to culture the cells and keep them alive.

That duty rested, according to Jiho Choi and Ok Jae Koo, on the shoulders of one man—Sun Jong Kim, the researcher at MizMedi who became a part of Hwang's lab because of his supposed expertise with cell culturing. Hwang's lab was adept at nuclear transfer, but as veterinary scientists, they were less familiar with culturing cells—keeping them alive in their pluripotent state with just the right media and growth factors.

Despite repeated attempts with hundreds of eggs, however, Kim could not get a human stem cell line to grow in culture. Then, says Koo, almost overnight the lines started to thrive. "Researchers from our lab also tried to isolate stem cells from the blastocyst," Koo tells me during a meeting in Seoul on the day Hwang was found guilty of fraud and embezzlement and given a suspended jail term of two years. "But we all failed. With Sun Jong, every time he tried he could get stem cells."

The key to Kim's success, everyone thought, was in the feeder layer he was using. Feeder cells are the living carpet that coats the bottom of a petri dish and mimics the natural environment that nurtures cells to grow. For human embryonic stem cells, this feeder layer was actually made up of mouse fibroblast cells—the same ones Martin Evans and Gail Martin had each first used to grow mouse cell lines—that secreted the right combination of growth factors, enzymes, and nutrients. Every morning, Kim would bring a special concoction of feeder cells from his lab at MizMedi to seed the human stem cells extracted by Hwang's team. "When Sun Jong prepared the feeder layer, almost every day it worked," said Koo. "We called him the god of isolating stem cells."

Within months, however, scientists were pointing out suspicious similarities between images published with Hwang's 2004 paper that implied the team had not generated as many stem cells as they claimed. After investigations by the university, it became clear that Hwang's stem cell lines had not been derived by nuclear transfer at all, but had indeed come from the more traditional source of IVF embryos—specifically, those from couples who had undergone IVF at MizMedi's infertility clinic. Kim, Koo suspects, may have brought feeder cells already containing stem cells from the IVF embryos into Hwang's lab and passed them off as the nuclear transfer lines.

Why the young researcher took such a risk, putting both his and Hwang's careers in jeopardy, still isn't clear, even years after the scandal. When I ask Koo about what may have motivated Kim, he pauses for a few minutes, then offers up his view of the expectations and the burden that his colleague must have felt to culture those stem cells. "In Korea, the professor is a very respected person. More than that, with professor Hwang, if he said something, it's the law of our lab. It may have been very stressful for Sun Jong that he should isolate stem cells from blastocysts. Already it was taking too much money and so many people related to the project to bring oocytes from the egg donors and doing nuclear transfer. And only Sun Jong could do that work on this very big project. Usually even one professor can be very stressful for one student. . . . But in his case, there were around ten professors, and Sun Jong was the only student."

Hwang still has not provided his version of exactly what happened in those months between 2004 and 2006, instituting a self-imposed media silence until, according to those who remain close to him, he proves himself right by repeating his experiments—for real this time. After being dismissed from his faculty position at Seoul National University in the wake of the scandal, he established a company, Suam Biotechnology Institute, to focus on animal cloning.

Despite the fraud involving the human cells, one scientific achievement still belongs to Hwang—cloning the first dog. As the allegations of scientific misconduct in his lab began to spread in late 2005, even this milestone, which was the lab's third in two years, was tainted with suspicion. But after an independent lab confirmed that the DNA from the donor hound and the cloned puppy, named Snuppy, were identical, Hwang was credited with the success.

Still, the damage to his reputation, at least in the scientific community, had been done. Hwang can no longer work with human embryonic stem cells in Korea, although he is rumored to be partnering with foreign collaborators to repeat his human cloning studies. His appeal is still such that despite the scandal, nearly all of his lab members, with the exception of only a handful of dissenters, left with him to work at Suam. Choi was among them, and he remains fiercely loyal to Hwang. They remain in regular e-mail contact, and before leaving to study in the United States, he even asked Hwang to preside over his wedding, an expression of the esteem with which he still holds his former mentor. "He is an honest man, a diligent man," says Choi. "He is not a cheat or a fraud; he would never do that kind of thing."

For the stem cell community, it meant only one thing: They were back to where they started. No one had yet been able to make a line of embryonic stem cells directly from a patient's own cells.

10

Safe Harbors

Stem cells were the furthest thing from the minds of the Estess sisters growing up in Rock Island, Illinois. In fact, aside from the occasional frog (and grasshopper and butterfly) dissection that mesmerized Valerie during summers in their playhouse, biologist and doctor were not high on the siblings' that's-what-I-want-to-do-when-I-grow-up list.

That was especially true for Jenifer, the third of the five siblings, who showed some acting talent in her high school debut. By then the family had moved east, to Harrison, New York, where the children's maternal grandparents lived.

But it wasn't the stage and its spotlight so much as backstage that captivated Jenifer. After graduating from New York University with a degree in theater, she landed, she thought, the perfect job. As producing director of Naked Angels, Jenifer did whatever had to be done to keep an off-Broadway theater group alive and acting—finding talent, nurturing it, raising money for it, and above all, putting on a good show. At some point during Jenifer's tenure, company members Marisa Tomei, Rob Morrow, and Gina Gershon walked across the Naked Angels stage, and Jenifer took pride in knowing that the troupe was earning a reputation as the place for launching talent. Even John F. Kennedy Jr. was on the board. "Some nights thinking about work, I was too excited to sleep," she wrote in her 2005 memoir, *Tales from the Bed*. "I was crazy in love with my job."

Too bad it didn't pay the rent. After five years, Jenifer tore herself away from Angels, eager to get on with her own Hollywood dream of producing films. She took a better paying—but less fulfilling—job and, Valerie tells me, "she was on the rise in New York. She really was the Carrie Bradshaw type—but cooler."

Jenifer, with a thick mane of dark hair and penetrating gray eyes, was living her dream in Manhattan, settled in the perfect apartment filled with carefully chosen tables and sofas and urban accoutrements; in charge of her life and her career; and convinced that her perfect partner was about to make his entrance from stage left—or stage right; it didn't really matter.

Like any good producer, she had her grand plan and perpetually saw the big picture. She never imagined that instead of putting those skills toward producing blockbuster films or plays, she would soon be orchestrating a medical miracle.

★ ★ ★

The first signs didn't really stop her. In fact, they were more annoying than anything else. They slowed her down and got in the way of her getting to where she needed to go and getting things done. It got harder to cover more than a city block without breaking into a sweat. Climbing stairs was more effort than mere exertion. Getting up out of particularly low-slung chairs was becoming a problem—sometimes an embarrassing one. And then there were the shakes. Inexplicably, out of the blue, Jenifer's legs would start jumping on their own. TAP, TAP, TAP, TAP, TAP.

Valerie and her younger sister Meredith attributed Jenifer's breathlessness to exhaustion, and the twitches, while odd, to stress. And so did Jenifer, at first. Who wouldn't be panting after juggling all the things she had on her plate—working full time, making plans for her growing production company, living the full-fledged I'll-take-Manhattan life? Why wouldn't her body be crying "uncle," desperate for a break? But the involuntary spasms were getting harder to ignore. "I can see your legs," Meredith would say. That was code that Jenifer's limbs had begun their macabre dance again, in a choreographed pattern of trembling and seizing that made sense only to the muscles and the neurons firing them.

In March 1997, the spasms finally lurched their way into a visit to the doctor. That was when the Estesses first heard the words *motor neuron disease*.

"When I heard motor, I immediately thought of going somewhere, of getting up and running out, and never stopping," Jenifer wrote. "The phrase itself—motor neuron disease—didn't conjure up anything specific."

That would soon change, as those words, combined with another innocuous-sounding jumble—amyotrophic lateral sclerosis, her ultimate diagnosis—quickly infiltrated the family's lives and came to define each one of Jenifer's siblings in her own way. "It became my reason to be," Valerie tells me of how ALS impacted her life.

It's been a dozen years since that fateful day when she and her sister first heard about neuromuscular disease. Valerie, a fit, athletic woman with a sheath of perfect blond hair and the same clear, penetrating eyes as her sister's, is now scientific research director of Project A.L.S., the non-profit organization she, Meredith, and Jenifer founded in 1998 in an effort to save Jenifer's life. Valerie is now well versed in the intricacies of neural degeneration. She has educated herself about ALS enough to sound like a pro, reeling off medical terms and describing the latest treatments with an articulate, matter-of-fact confidence rivaling that of any scientist. "ALS is a targeted destruction of motor neurons, which are large brain cells that reside mostly in the spinal cord but also in some parts of the brain," she begins. "No one knows why motor neurons are selected for this destruction, but they are. As the motor neurons die, so goes a person's ability to speak, breathe, swallow—any voluntary movement. Imagine these neurons being progressively destroyed over the two- to five-year course of this disease. That's the prognosis. It's not very pretty. And there are no interventions that can make a difference in the disease."

Valerie should know. She witnessed just how "not pretty" the disease can be. Excruciating muscle by excruciating muscle, it robbed her sister of her ability to walk, talk, smile, laugh, cry, breathe, and, ultimately, live. Valerie has chosen as our first meeting place a small, stark room off the Project A.L.S./Jenifer Estess Laboratory on the Columbia University Medical Center campus, in New York City. This five-thousand-square-foot fourth-floor lab is Jenifer's legacy. In 2003, just before Christmas, Jenifer Estess died of ALS. She closed those bright gray eyes never knowing how her personal struggle with the disease that claimed the life of Lou Gehrig would become the inspiration for the first stem cell–based research lab for the disease, or how her dedication to promoting research into ALS would push it to the top of the list of conditions to which stem cell therapies might be applied first.

★ ★ ★

Jenifer would have laughed her big, hard-to-ignore howl over the idea that her name and image would appear on the door of a state-of-the-art molecular biology lab. And she would not have taken seriously any suggestion that her story would move millions to give millions in the name of research for a disease that she had never heard of.

She was the creative one, practicing monologues, losing herself in movies—*Funny Girl*, with Barbra Streisand, was a favorite—always imagining, transporting herself somewhere else as someone else. Valerie was the boss; when things got rough for the Estess clan, it was she who rounded up the troops and rallied them to pull through. After their parents divorced when Valerie was fifteen, their mother "took a powder," struggling for two years with the dissolution of her marriage by essentially checking out of her family's life and remaining ensconced in bed for days at a time. Their eldest sister, Alison, left for college, while their younger brother, Noah, who had epilepsy, was sent to live with their father. That left Valerie, Jenifer, and Meredith. And Valerie quickly became coach, organizing everything from football-throwing lessons to dissection demonstrations to homework sessions and lunches. It was Valerie who took over when things slid south.

So it was no different that March. The diagnosis began a slow but inexorable shift in the axis of Jenifer's world—ever so slightly at first, but ultimately into an impossibly contorted twist that seemed to gain momentum with every day. Hoping to keep her life as "normal" as possible, Jenifer hired Meredith to work with her and cover for her if her legs started to fail her. Then her beloved apartment on Seventy-first street, that urban oasis that both defined and sustained her, became an impossible-to-reach prize with every venture outside. And finally, when too many of the muscles in her arms and legs had succumbed to ALS's march up her spinal cord, there was her bed, a queen-size sanctuary that anchored her rapidly spinning, out-of-control world. Without even thinking, Valerie says, she, Meredith, and Jenifer agreed together that they were "gonna get out of this." Somewhere, someone, they figured, had an answer to treating ALS. How could they not? Some hub of ALS research would be their golden goose; surely the scientists in that lab would have something in their beakers and petri dishes that could help Jenifer.

But despite call after call and trip after trip, Valerie, who had naturally assumed the role of family ALS authority, found that no such hub existed. "We found no one leading the charge," she tells me. "We went to places like the Muscular Dystrophy Association and the ALS Association; we rounded up the usual suspects. And we found organizations funding mostly patient services, helping people adjust to the disease by helping them to find wheelchairs, walkers, assistive devices, and other ways of coping—all of which are crucial and key. But they were spending smaller parts of their budgets on actual scientific research. Which was shocking to us."

She finally decided to take the train with Meredith to Bethesda, in order to debrief, in her trademark take-charge way, the National Institutes of Health on the latest in ALS research. As the nation's leading site of basic medical research, it was a logical place to start. But it was early 1998, and Jamie Thomson and John Gearhart had yet to announce their respective breakthroughs with stem cells. When it came to brain disorders—the so-called neurodegenerative diseases that include Alzheimer's,

Parkinson's, and Huntington's—scientists were pursuing all kinds of interesting therapeutic leads. But the brain and the nervous system being what they are, those strategies weren't steering them toward any real treatments or cures anytime soon. Much of the research at the time was still at the very basic science level, too far removed from anything resembling a treatment ready for testing in human patients. But that's what the Esteses were looking for—anything that would help their sister.

The true promise, most of these experts would eventually realize, lay in exploiting stem cells, both as learning tools to shine a much needed light on the mystery of what was causing nerve cells like Jenifer's motor neurons to suddenly stop working, as well as to hopefully spawn new treatments.

Valerie continued crisscrossing the country, often with Meredith, leaving Jenifer home; by then she had moved into a room in Meredith's house and into a wheelchair.

Still in coach mode, Valerie immersed herself in the ALS literature, homeschooling herself in the intricacies of nerve growth and development and the latest theories about restoring their function. In 1999, she came across the work of Jeff Rothstein, an ALS researcher at the Johns Hopkins Hospital, in Baltimore. If doctors have a reputation for being too clinical and treating patients as objects, Rothstein could have been that model. Young, matter-of-fact, and even brusque, he tore through an ALS 101 lecture with Jenifer, Valerie, Meredith, and their mother after they squeezed into his tiny examining room on Wolfe Street.

"There's no question you have ALS," he told them after running Jenifer through a gauntlet of physical exercises involving pulleys. "Your motor neurons are dying and they'll never be replaced."

That wasn't news to the family. What was more significant about that visit was Jenifer's question in response to the grim news: "We can replace damaged hearts and kidneys, so why not motor neurons?"

Why not motor neurons?

Such a simple question, and a logical one. If ALS is a disease in which the motor neurons are dying, why not replace them with perkier, more active ones? By then, Thomson had shown that embryonic stem cells could be harvested from IVF embryos. Surely these stem cells could be guided to become, among other things, the motor neurons that were waving the white flag in Jenifer's spinal cord?

"Cell transplants? That's science fiction," Rothstein responded, quickly extinguishing the spark of hope flickering in Jenifer's query. He was dismissive of the preliminary nature of stem cell work and the years it would take before it would even be clear if they could serve as viable sources of new cells for transplant. There was a finality with which he delivered his opinion that was intended to deter Jenifer from having any fantasies about being cured with stem cells.

But neither Jenifer nor her sisters could be waved off that easily. "Her question really got us going on this idea, and we would not let up," says Valerie.

Why not motor neurons?

Several months later, they got to ask the question again, this time of Evan Snyder, a neurologist working at Children's Hospital Boston on shiverer mice. These mice are bred with a genetic disorder that short-circuits the connections between their central nervous systems and their muscles, causing them, as their name suggests, to shudder involuntarily. Snyder had become fascinated with the potential of stem cell transplants, and, like Jenifer, wondered why he couldn't replace the damaged neurons in his shiverer mice. So he injected a pool of neural stem cells into the spinal cords of several mice and, as he reported in the journal *Proceedings of the National Academy of Sciences*, some of them stopped shivering.

"That was good enough for a kid like me," says Valerie. If it worked in the shiverer mice, then maybe it could work in mice with ALS. And maybe that would mean it could work in Jenifer.

After reading Snyder's paper, Valerie called him and convinced him to try the same thing with an ALS mouse. "We just crossed our fingers," she says.

The result: failure. "Abject failure," Valerie tells me, stressing how fantastically they had miscalculated the complexities of the mammalian nervous system. "We were disappointed; we were crushed."

And yet, there was something about stem cells that the Estesses just couldn't shake. Call it a feeling; call it inspiration or even desperation. "From that moment, we knew it was 'There's gold in them thar hills' kind of thing, says Valerie. "There was something about stem cells—at least in theory, if not in initial practice—in those first experiments, and we were determined to figure out what we might be able to mine these for."

★ ★ ★

A good idea, but like the first gold prospecters, scientists had no guide through the hills of early development, and no real understanding of what stem cells were yet, much less how they could be manipulated to mature into something as complex as a motor neuron.

By November 1998, as the Estess clan was gathering for the holidays, Valerie was characteristically impatient. Watching Jenifer move from a cane to braces and finally to a wheelchair, she could see the march of the disease every day. And she was getting edgy about their chances of finding someone—anyone—who could help her sister halt the death of her motor neurons. Time, she felt, was running out.

What was frustrating to Valerie was the way in which basic research into ALS was conducted. She saw brilliant scientists working in isolation, like talented orchestra musicians each blaring their instruments to their own beat and in separate keys. The end result was a biological cacophony rather than a coherent melody.

"I remember saying, 'Let's lock up the smartest scientists in a room, put a gun to their heads until they come up with a plan of action, kind of like the Manhattan Project,'" says Valerie. "They created the atom bomb—why not a treatment for ALS?"

So Project A.L.S. was born, in Meredith's living room, just before Thanksgiving. The idea was simple: Raise money, which the Estess sisters—Jenifer, with her production experience in off-Broadway New York; Valerie, who worked in corporate advertising; and Meredith, whose expertise was in fashion merchandising—were convinced they could do easily, and funnel it to the leading neuroscientists in the country.

The money would come with very tenacious strings, however: the sisters' determination for a treatment for Jenifer. Those strings would only get pulled more with every year, owing primarily to Valerie's insistence that every researcher remember the purpose of their studies. "My job is to keep the scientists on task," says Valerie, with her typical drive for results. "It's really important for everyone to understand that we need to meet certain milestones. When we fund a project, it's not a grant so much as you are getting paid to do this."

After the failure of Snyder's experiments, even Valerie recognized the need to step back and reassess exactly what stem cells could do and how they might best be exploited in treating ALS. Ever the adept recruiter, she brought Thomas Jessell, a neuroscientist at Columbia University, into the Project A.L.S. family. "We soon learned that it was not going to be as straightforward as putting cells into paralyzed animals and then having those animals get up and dance the watusi," says Valerie.

It was Jessell, a sharp and inquisitive Brit, who put the brakes on the Estesses' drive to transplant stem cells, redirecting the project's energies toward answering some more fundamental questions about normal neural development first. "If one understands at a fundamental level the way a neural circuit assembles itself for function, then you have a better chance of intervening and manipulating that circuit under conditions when things start to go wrong," he says of his rationale. The key, he thought, would lie in understanding as much as possible about the normal progression of neuron formation and then comparing this knowledge to that gleaned from watching neurons generated from stem cells as they grew in a dish.

Jessell had already spent the better part of two decades pursuing this systematic strategy—not with stem cells, but with methodical investigations of the various genes and molecules that together converge to dictate first which nerves out of the body's five thousand to ten thousand classes become motor neurons, the specialized nerves that connect to muscles, and then how these motor neurons link up with their designated muscle mates among the six hundred different muscle groups in the body. And he had made considerable headway, identifying a family of genes that seemed to regulate this process.

"Then I was told to stop wasting my time and start thinking about how to apply some of these basic sets of questions and observations to something that actually might be useful to someone at some point," he says, laughing, of his initial meeting with Valerie.

Jessell wasn't a fan of the stem cell approach at first. In fact, the focus on using embryonic stem cells to generate a source of new tissue to transplant into patients made him wary. "I was a little bit reluctant to get involved with the field, because the biology and science weren't so far advanced, and were open to conjectures and hypotheses that didn't have much rigorous scientific basis."

A bright postdoctoral fellow in the lab, however, convinced him otherwise. Hynek Wichterle was completing his Ph.D. at the Rockefeller University, studying nerve development in the brain, when the first human embryonic stem cells were isolated. He became intrigued with the idea of using the stem cells to make new neurons, but not for any clinical purpose—his work was deeply embedded in the intricate cellular processes that drove neural precursors to form the five thousand to ten thousand different subtypes of nerves. He simply wanted to tease out all the steps it would take to bring a motor neuron all the way from an embryonic stem cell to its final specialized state. After two attempts, he finally convinced Jessell to let him try. "I thought that was a far-fetched idea, because if motor neurons form as one in one thousand possibilities from a neural progenitor, then there is even less potential of forming a motor neuron from an embryonic stem cell, more like one in ten thousand or one in fifty thousand," says Jessell. He gave Wichterle six months. If the postdoc wasn't successful, then he would have to move on to another project.

In short order, Wichterle proved to Jessell that his skepticism was unfounded. With the deft application of the two decades of accumulated work on nerve development that his mentor had amassed, Wichterle managed to march mouse embryonic stem cells through their various developmental stages from neural progenitors into motor neurons. What's more, he also introduced his lab-made cells into the spinal cords of chick embryos and showed that they adapted and functioned just as well as normally formed motor neurons in enervating limbs.

Maybe stem cells, Jessell had to admit, were not so bad after all.

★ ★ ★

Wichterle's success was a milestone in the nascent stem cell field. It was the first demonstration that a specific type of cell could be coaxed to develop from an embryonic stem cell. Yes, it was in mice, but if it was possible in mice, it would likely be possible with human cells. And indeed it was. Others, including Wichterle, repeated the process using human ES cells, although the efficiency was much lower. It also took longer for the human motor neurons to show themselves, something that initially confused Wichterle. "I tried the exact same protocol I used for the mouse, which took seven days, and it failed," he says. He extended the study to fourteen days. Still nothing, Then twenty-one days. No motor neurons. It would take about a month for the human nerves to develop, which highlighted for Wichterle how critical using human cells would be, especially to understand differentiation from embryonic stem cells. As useful as mouse models are, they are, as scientists learn over and over again, not always good stand-ins for man.

Wichterle's next steps were to put Jessell's two decades of work to the test, using the recipes that Jessell had written for developing the various types of motor neurons to figure out which ones were dying in a disease like ALS and which ones were not. "Not all motor neurons are dying equally fast in ALS," says Jessell. "So the rationale is to make motor neurons that never die in the disease, and to make motor neurons that are extremely prone to death, and understand what the differences between them are. Then you have a rational approach to understanding how to think about a treatment intervention."

In order to do that, however, they would need human stem cell lines. And that's where Project A.L.S. could help. In 2006, Valerie and

Meredith opened the laboratory named after their sister, hoping it would become a safe harbor for anyone who wanted to work with human ES cells but could not because of federal restrictions. They were not making a political statement per se, but at the same time they were loath to see anyone *not* pursue a line of research simply because they couldn't. "They came with the idea, and were willing to fund this lab to work on human motor neurons and solve all these issues that bureaucratically made it a very unpleasant idea to start work on human cells in my own lab at Columbia," Wichterle tells me. He admits that were it not for Project A.L.S. providing the space and the funding for expanding his studies into human cells, he would probably not be doing them at all. "If there was not the option of this lab, I would not be involved in the work on human motor neurons; I would not be involved in the more directed work of trying to find some clinical application for motor neurons," he confesses. "I would probably be asking questions about basic development, like what underlies motor neuron diversity."

★ ★ ★

The Estess sisters were certainly not alone. With the government funding only a limited number of human embryonic stem cell studies, patients—and researchers—were finding other ways to get the research done. It was not an ideal situation—private philanthropy, while powerful, can never match the deep coffers of the government when it comes to supporting biomedical research—but it was something.

And, as Valerie Estess and her scientific brain trust were discovering, there was an advantage to being small and sleek. They could move more quickly than a behemoth like the NIH to pick up new ideas, test them, and drop them if they proved unworthy. There was a nimbleness to their approach that was beginning to appeal to researchers accustomed to the inertia of bureaucracy.

And that was the idea behind the New York Stem Cell Foundation (NYSCF).

★ ★ ★

On one of those humid August evenings in New York City in 2005, an impressive group of scientists and patient advocates were assembling in Manhattan at the request of Susan Solomon. Dark-haired and motherly, Solomon, a corporate attorney who describes herself as more comfortable behind the scenes, was not accustomed to taking charge in this way, but she had finally had enough. Raising a son with type 1 diabetes and caring for a mother in the last stages of Alzheimer's disease, she felt the government's restrictive stem cell policy was unacceptable. But it wasn't until she received a call from Harold Varmus that she was spurred into action.

Varmus, who was by then president of Memorial Sloan-Kettering Cancer Center, had not forgotten the stem cell fight that had occupied his last days in Washington. And three years into Bush's policy, he was perplexed by the lack of hue and cry he had expected to hear from the patient advocates. Solomon, whom Varmus had gotten to know after his move to Manhattan, was an active member of JDRF. Where, he asked her, were all the patient groups? Why weren't they standing up for stem cell science?

"The large disease advocacy organizations were using a political calculus," Solomon says as we talk in her offices near Lincoln Center. "They were triaging. They felt it was too early to come out in favor of stem cell research and risk the almost certain ire of the administration They thought they could serve their constituents better by staying silent, staying on the sidelines, and pushing for their own NIH allocations. And that was too bad."

The more Solomon thought about it, the more she realized, as Melton had, that the burden of taking the promise of stem cell science and turning it into cures would lie with private groups. But it couldn't be just any non–federally funded groups—they had to be willing to make stem cell research a priority, even if it proved unpopular and financially risky. It had to be grassroots.

Recruiting Varmus and then Doug Melton, as well as Tom Jessell and Paul Nurse, a Nobel laureate from the Rockefeller University, and a handful of the field's leading experts to the medical advisory board, Solomon rounded out the foundation's think tank with prominent patient advocates, including Chuck Close, the photo-realist who had suffered a seizure that left him partially paralyzed. An anonymous donation of one million dollars got the nonprofit on its feet.

And at that muggy dinner, the board checked off the top three critical needs in the field: a pipeline of young researchers to ensure a next generation of scientists who would continue the work that pioneers like Melton and Jessell had started; a community to nurture stem cell scientists, feeding them both intellectually and psychically; and a safe haven lab where they were welcome to and not restricted from conducting work on human embryonic stem cells.

It was Nurse who gave NYSCF its first and most important mission that evening—to fund training grants for young scientists who might be scared off from stem cell studies by the government's restrictive policies. "He said we have already lost a generation and people don't realize it," Solomon says. Graduate students and postdoctoral fellows were wary of entering a field in which the largest grant-giving institution had shrunk its pipeline down to a trickle. Many were being steered toward neuroscience and advised that stem cells were simply too volatile.

With the scientific stem cell community still a fractured group, Solomon also saw an opportunity for the newly formed foundation to serve as facilitator for bringing researchers together to share their findings and exchange ideas. With so few scientists able to raise the private funding necessary to support work on human embryonic stem cells, there was little opportunity for them to gather and learn from one another. So the following year, NYSCF launched the first of its annual research conferences, themed Curing Disease from Lab Bench to Bedside. Solomon, like the Estesses, did not want the scientists to forget why they were there.

That year, the foundation also named its first NYSCF fellow, a Rockefeller University postdoc, for her proposal to study embryonic stem cells in hair cells and cancers of the skin. The foundation now funds seventeen fellows in the New York area, who were attracted to NYSCF not only for the financial backing but for something far more important to a budding scientist—the access it granted to a state-of-the-art lab on the Columbia University campus, just down the hall from Project A.L.S.

★ ★ ★

The NYSCF lab was a refuge for stem cell studies, and indeed its first project in 2006 was Kevin Eggan and Doug Melton's program to generate human ES cells using nuclear transfer. At the time, Massachusetts governor Mitt Romney was opposed to all forms of cloning, so to ensure he did not run afoul of state laws, Eggan performed his first cloning attempts with human eggs at the NYSCF lab. Eggan has since passed on his expertise to a postdoc who is an NYSCF fellow and carries on the legacy of perfecting the technique with human cells. In 2009, the New York state legislature passed a law allowing researchers to compensate women for donating eggs for research, and Solomon is hopeful that the move will spur enough volunteers to donate so the research can continue.

The lab remains one of the few in the country where scientists—both those in training and those further along in their careers—can learn this critical technique. And with the academic institutes nervous about cross-pollinating resources from federal and non-federal entities, the lab was a welcome option. "We act like Switzerland," says Scott Noggle, director of the laboratory. "Our mission is to collaborate and provide a safe haven as an independent research foundation."

As Melton's did, Noggle's group also derives its own human embryonic stem cell lines from excess IVF embryos, and distributes them to labs that need them. And as the science evolves, the foundation is tagging along for the ride. As knowledge about how to manipulate embryonic stem cells and turn them into specialized cells matures, the needs of the researchers will grow as well. Already, for example, as CIRM is finding out, scientists are turning their attention to using stem cell lines to screen for new drugs, something NYSCF is planning for as well, by adding the ability to survey thousands of compounds in the same way start-up bio-tech companies would.

NYSCF may have been born out of political need, says Solomon, but it will leave a scientific footprint that will far outlast any policy. "You start something because you need to have a safe haven lab because of political issues," she says. "But then you end up aggregating expertise that becomes its own sticky community."

That's also true for the previously disparate scientists Valerie Estess managed to "lock up" to work on ALS. Working together, the neu-rologists and the stem cell scientists have already exposed a new culprit that may be responsible for picking off motor neurons in the disease. The neurons, it turns out, may not be faulty themselves but merely the victims of the toxic by-products of other nearby nerve cells. It's a clue pointing toward a potential treatment, and while it may or may not yield anything useful, it is, says Estess, a step forward. "And Jenifer taught me that, that you always have to move forward."

11

A Jolt from Japan

The city of Kyoto holds a special place in Japanese history. As the twelve-hundred-year-old second capital of the country, it is home to the original imperial palace, and is the cultural heart of all things Japan, from the ritualistic tea ceremony to artisanal skills in bamboo craft and textiles. Serene ryokans, the elegant, paper-walled domiciles of the ancient court that once dominated Kyoto society, preserve a lifestyle devoted to the pursuit of peace and harmony with nature. Stone gardens, the Zen oases of carefully raked sand, still attract millions of visitors, both native and otherwise, in search of spiritual meaning and serenity.

Flowing down from the Kitayama Mountains that border Kyoto's northern limits, the Kamo-gawa River is a pure and narrow conduit into the center of the city. With the precision of a samurai's sword, it slices through Kyoto, creating unequal halves of the urban and the ancient. The riverbed is well worn, etched into the earth by centuries of gushing waters that seemed to flow, by design, alongside the palace just to quench the thirst of the King and his imperial party.

The riverbanks are renowned for inspiring meditation, and the cherry tree–lined Path of Philosophy is named for the countless thinkers of Japan who wandered its riverside borders. In modern Kyoto, the canals also serve to demarcate past and present: On the western banks of the river sits the Kyoto Imperial Palace, while just opposite sprawls Kyoto University, with a glistening new structure boasting all of the latest innovations in high-tech science. If the palace signifies Kyoto's storied past, then the university symbolizes its cutting-edge future.

It was here, in a second-floor lab on the school's campus, that the single most important breakthrough in stem cell science occurred. In 2006, Shinya Yamanaka turned back cellular time.

Bespectacled and lean, the modest Yamanaka betrays none of the arrogance of a scientist or even the swagger of a surgeon, which is how he began his career. He is a modern blend of the culturally polite Japanese and casually friendly American, an artifact of his days as a fellow in California, where he still maintains a lab. He likes to disarm new acquaintances— including the thousands that now crowd every presentation he gives at scientific conferences—with a few laughs. At the International Society for Stem Cell Research (ISSCR) conference in Barcelona, Spain, in 2009, Yamanaka bookended his presentation with two slides of his lab members. The photos are practically a prerequisite for these talks, a way for senior investigators to acknowledge and recognize the efforts of the students and postdocs who are the lifeblood of their labs. Yamanaka, however, can't resist the opportunity to lighten the mood at the generally serious sessions. Breaking with tradition, he opens with a group photo, pointing to a male student decked out in the elaborate costume of a geisha; in the final slide, the same student appears again, this time in a blue bunny costume. "That boy is still strange," he says, to a room of appreciative laughs.

Before I enter Yamanaka's office, on the second floor of the Institute for Frontier Medical Sciences, I am greeted by a makeshift sign in Japanese, printed from a computer and taped between two tensile barriers like the kind you see at airports to corral wayward crowds. THE LABORATORIES ARE CLOSED TO ALL UNAUTHORIZED PEOPLE, the sign warns. THANK YOU FOR KEEPING OFF.

It's a consequence of the enormous fame and recognition that the introverted scientist now commands, an attention that he is only reluctantly beginning to accept. Outside are more signs of his influence: Steps away, a gigantic construction tarp shrouds what will be Yamanaka's new home—the five-story Center for iPS Cell Research and Application (CiRA), a facility dedicated to nurturing the remarkable stem cells that Yamanaka has created.

Inside his moderately sized office, two huge arrangements of orchids spill their shower of white flowers over a conference table. Just two weeks ago, he was recognized with a Lasker Award, a coveted prize honoring scientific achievement that many feel is a prelude to the Nobel Prize, and the flowers are a congratulatory gesture from well-wishers. The room is very professional, and functional; aside from the flowers, there are few signs of personal effects that betray anything of the personality of the scientist who occupies it. Taking up the most space in the office is the long conference table that turns the small area into a meeting room for lab members.

It was around this table, and in this room, that Yamanaka worked out the details of how to reprogram adult cells. The technique, in retrospect, is at once simple and brilliant, and represented a true leap of scientific faith. But, as with many such breakthroughs, scientific and otherwise, it almost didn't happen.

★ ★ ★

"I was very good at doing surgery on mice, but not on humans," Yamanaka confesses. "I don't know why."

"I'm sure that's not true," I say.

"It is true," he says earnestly. "I found myself not good at surgery."

Yamanaka is being characteristically modest, and is uneasy because the topic of conversation is himself. Closing his eyes every once in a while in thought, however, Yamanaka is engaging and often funny in his choice of words to describe the roundabout journey that brought him to stem cells.

It was Yamanaka's father who encouraged him to become a physician. So in 1987, the dutiful son graduated from Kobe University with a medical degree. He began his residency at Osaka National Hospital in orthopedic surgery. These doctors are the well-known commandos of the operating arena, and many take a disproportionate amount of pride in the tools of their trade—drills, saws, sanders, and the other home improvement–worthy implements that they use to work away at hardened bone and cartilage, or to smooth out misaligned or damaged joints. After two years, the reserved Yamanaka found himself turned off by the bravado of the operating room and more drawn to the discipline and intellectual rigor of basic science.

He enrolled as a Ph.D. student in pharmacology at the Osaka City University Graduate School of Medicine. There, in Katsuyuki Miura's lab, Yamanaka was introduced to the magic of transgenic mice—rodents genetically modified to contain and express human genes in their genome. Working on manipulating genes, Yamanaka was at home.

Armed with his Ph.D., Yamanaka showered university labs in the United States from coast to coast with his application to study mouse molecular genetics. Almost none were interested or ever even replied. "It was very natural that a failed surgeon with little experience in molecular biology had a hard time finding a position," he wrote in a commentary in *Nature Medicine* of his attempts to find a fellowship.

The one exception was Thomas Innerarity, at the Gladstone Institute of Cardiovascular Disease, in San Francisco. In the first of what would become a series of professional rescues, Innerarity, who was studying fat metabolism, faxed the Japanese scientist back; he was interested in hiring him.

Thrilled, Yamanaka packed up his belongings and moved his wife and two daughters from Japan to California. Gladstone was a haven. The newly created institute, housed in a fancy glass and concrete complex in San Francisco's Mission Bay, formed the core of what would become the city's science hub. Nestled along the southeastern edge of the city, nearly every lab or office on the eastern side of the building commands a stunning view of San Francisco Bay. California was a far cry from the restrained culture of Japan, and Yamanaka found the scientific environment at Gladstone fertile ground for his scientific curiosity. (Occasionally the legacy of his stay in the Golden State betrays itself in his speech, by his liberal use of the word *like*.)

At the time, however, Yamanaka wasn't even interested in stem cells. That would come later. What preoccupied him then was a gene involved in cholesterol metabolism. Innerarity was studying a protein, apolipoprotein B (apo B), which was the primary component of bad cholesterol, the low-density lipoproteins (LDL) that stubbornly coat artery walls and contribute to atherosclerosis and heart disease. A single gene, it turned out, produces two forms of this protein—in one version, the entire protein is generated; in the other, a truncated form is produced. This shorter protein, Yamanaka speculated, could lead to a lightening of the cholesterol load by helping to generate fewer LDL particles.

The key, he soon learned, was an enzyme that was responsible for cutting the protein as it unspooled from the nucleus of the cell where it was made. Yamanaka found that in the liver, the full-length protein was produced, while in the intestine, the enzyme worked diligently to snip up the protein into its smaller, less dangerous form. He theorized that coaxing the liver, where most of the cholesterol is produced, to churn out more of the enzyme would lead to more of the shorter apo B being made and a lower risk of heart disease.

Excited by the potential clinical utility of his work, Yamanaka eagerly tested his theory in a group of rabbits and mice. He generated transgenic strains of the animals containing the human gene coding for the enzyme, and checked the blood cholesterol levels of the animals. To his relief, their LDL levels were significantly lower than those of normal animals without the gene.

Very quickly, however, Yamanaka realized that something was wrong with the mice. "It turned out that the gene actually functions as a [cancer-causing] oncogene," he says. "So many of those transgenic mice developed tumors in their liver." It was a crushing blow to Yamanaka, who had hoped the gene would lead to some kind of therapy for heart patients.

Disappointed but undeterred, Yamanaka next turned his attention to asking why—why did the process of editing the protein lead to the formation of tumors? What other genes were turned on or off when the gene for that enzyme was over-expressed in the liver?

Within months, he had homed in on an answer: another gene, which he dubbed NAT1, whose protein was prematurely clipped. In its shortened form, this protein led liver cells to grow uncontrollably into tumors. Yamanaka knew that to learn more he would need mice that had been bred without the gene. Only by studying these mice and whatever physiological, metabolic, or physical abnormalities they developed could he understand the full spectrum of the processes, tissues, and organs that were affected by the protein made by that gene.

And such studies were exactly what mouse embryonic stem cells were made to do. As familiar as he was with mice, however, Yamanaka had no experience with stem cells.

But down the hall, his colleague Bob Farese did. Affable, with a perpetual California tan, Farese looks as if he might be more comfortable bobbing on a surfboard off the Pacific Coast than hunkered over a microscope in a molecular biology lab. As a postdoctoral fellow in biochemistry at Gladstone, he was, like Yamanaka, interested in lipid and energy metabolism related to obesity and diabetes. He was also a whiz at growing up embryonic stem cells. "Making stem cells from a certain strain of mouse is not that hard once you know what you're doing," Farese says. "Everybody works on mouse embryonic stem cells from that one particular strain. It worked the first time I ever did it."

Learning everything he could from Farese, Yamanaka began to see the power of stem cells as a platform for systematically studying the effects of genes. The simplicity of the technology appealed to his love of precision—he could generate stem cells, turn specific genes on or off, and observe how the changes affected the development of the mice made from them. Yamanaka didn't know it then, but he was setting the stage for what would become his pioneering work in reprogramming a fully mature cell.

When his three-year fellowship was over, Yamanaka returned to Osaka City University in 1996, landing this time in the Department of Pharmacology as a professor. He didn't come alone, however. As souvenirs he brought back with him NAT1 and three chimeric mice bred with the NAT1 gene. His immediate goal was to find out which genes were affected by NAT1—if NAT1 were missing, he wondered, what would happen to the mouse?

But in a department dedicated to elucidating the potential of new drugs to treat disease, this mouse work was sorely out of place. With little intellectual support, and even less financial support in the form of scientific grants for his experiments, Yamanaka could not afford to hire anyone to maintain the animals that were so critical to his experiments. The responsibility fell to him. So each week he cleaned and changed the mouse cages by himself—a task unheard of for almost any professor in any university anywhere in the world. Yamanaka was fully aware, however, that it was the price he had to pay for continuing his work—and he was therefore willing to do it.

He was paying another price as well, in the lack of a like-minded community of biologists such as the one he had enjoyed in San Francisco. "Only a few scientists at the institute [in Osaka] understood the importance of mouse embryonic stem cells," Yamanaka tells me. "They thought the project was too basic from the point of view of medical research. So I had to maintain many, many mice all by myself. And I had a hard time getting funding."

Ironically, just as he was hitting an emotional low, he was reaching a professional high. The NAT1 gene he had isolated, while of little interest to the pharmacology field, was actually an important factor in early mouse development. When Yamanaka generated mouse embryonic cells without the gene, the cells grew normally and expressed all of the genes that they should, except that they never completed their development. They never went on to mature and differentiate into specific types of cells. Instead, they remained stuck in an undifferentiated state. They remained embryonic stem cells.

It was a key finding, and one that ignited Yamanaka's interest in stem cells and the factors that drove their "stemness." But he seemed to be the only researcher in Japan who saw the potential in NAT1. As a postdoc in the States, Yamanaka had been scientifically prolific, publishing at least one paper each year. But his submission on NAT1 was rejected by journal after journal in Japan. "After going back to Japan, I thought it would be very easy to be accepted, but I learned it was actually very difficult," he says.

The combination of rejection and professional isolation devastated Yamanaka. The culture shock was overwhelming, too. After the satisfaction of being engaged in and contributing to the scientific community at Gladstone around the growing field of stem cells, he no longer had any colleagues to nurture and inspire his passion. "I got almost depressed," he admits, closing his eyes for a few seconds as he recalls his frustration. "I was almost about to quit science. I was thinking seriously of going back to the clinics as an orthopedic surgeon." He opens his eyes and adds, "I kind of died once, as a scientist."

Then came Jamie Thomson's breakthrough.

★ ★ ★

News of Thomson's and Gearhart's successes in cultivating human versions of the cells that Yamanaka was working with provided the lifeline that he desperately needed. Convinced that these ES cells would be tremendously important for medicine, the reinvigorated Yamanaka began looking for someplace, *any*place in Japan where he might find like-minded scientists to support his stem cell studies.

If he had died in Osaka, he was reborn in Nara. After applying at the Nara Institute of Science and Technology, a twenty-mile drive due east, Yamanaka had the chance to head his own lab for the first time. With this new lease on his professional life, he was not about to squander the opportunity. "Because I almost died once as a scientist, I thought I should do something very difficult and risky," he says, a slow smile spreading across his face. "I was ready to die anytime as a scientist again. That was a reason I started this—we knew this was a very risky project."

That project was the remarkable experiment that ultimately yielded iPS cells—the groundbreaking induced pluripotent stem cells that act and look like embryonic stem cells and are made not from embryos but from adult skin cells. Most of the researchers in the field were focused on manipulating embryonic cells to become something else. The potential of these versatile wonders of biology, after all, was in convincing them to become a fountain of youth for the body's damaged and diseased cells. The medical payoff lay in finding a way to reliably and safely generate, say, heart muscle to replace those scarred and lost to a heart attack, or beta cells in the pancreas to take over for those no longer able to produce insulin in diabetics, or dopamine-producing neurons in the brain to pick up for those that no longer connect the desire to move to the activity of a muscle.

Yamanaka thought otherwise.

"At that time, many people were trying to induce differentiation in human or mouse embryonic stem cells," he explains. "I thought I should do something opposite—because the field was very competitive already. So instead of doing differentiation, I thought I should do de-differentiation. Instead of differentiation of embryonic stem cells, I thought I should make embryonic stem cells from somatic cells, from already differentiated cells. The direction was totally opposite."

His contrarian approach wasn't simply a matter of obstinacy. After Thomson's experiments, it quickly became clear that because of the source of his stem cells—human embryos—his procedure would pose several major challenges when it came to generating a patient-ready therapy from them. Biologically, because they had originated from embryos, embryonic stem cells could not be immunologically matched to every potential patient; as with organs, the cells contained immunological identities that would render them foreign in anyone else but the donor from which they came.

And ethically, of course, there was always the problem of using human embryos for research. In Japan, research on embryos required a government license, which was granted only after an experiment passed a strict two-tiered approval process. Keenly aware of the debate boiling in the United States as well, Yamanaka saw clearly that such political concerns over research on embryos would only intensify.

As if to solidify his worries, he also had a personal encounter that convinced him that there had to be another way to generate stem cells. During a chance visit to a friend who worked at a fertility clinic, Yamanaka sat down at a microscope and peered down onto a tiny human embryo. By then he was the father of two young girls, and it occurred to him that each of his daughters had begun as a tiny ball of cells similar to the one under the lens. "I thought, 'We can't keep destroying embryos for research,'" he says.

So after much thought, Yamanaka steered his lab—and his career— toward what he saw as the next generation of embryonic stem cells.

★ ★ ★

Fueling Yamanaka's confidence were two critical studies. Soon after earning his M.D., he had come across a paper describing a fascinating mutation in flies. In these flies, a master gene—one that turns other genes on and off, much like a conductor cuing and silencing his musicians—is normally turned on in the insect's thorax but turned off in its head. An aberration in its function, however, led to the gene being activated in the head. When that happened, the flies remarkably sprouted a pair of legs instead of antennae above the eyes. What that meant was that a single factor controlled developmental processes as complex as the entire suite of steps needed to generate a fly's leg—everything from the chitin that it is made from to the proper articulation of the joints.

To Yamanaka, whose new focus was on reversing the developmental process, this was a clue. "From that drastic mutant I knew only a single transcription factor can make a huge difference," he says.

For even more assurance, Yamanaka found another intriguing example of rewritten development. In 1987, Robert Davis and his colleagues at the Fred Hutchinson Cancer Research Center had reported an astounding finding about a protein called MyoD. MyoD regulates muscle development, orchestrating the transition of uncommitted mesoderm cells into the specialized fibers that make up different muscle types. The gene coding for MyoD was so good at its job, in fact, that Davis could simply splice the gene into the genome of mouse embryonic skin cells and transform them into stable populations of muscle cell precursors. MyoD is normally not present in newborn mouse tissue, and is only expressed in skeletal muscle, the fibers that are responsible for the voluntary work of contracting and stretching when we want to move. Left alone in the culture dish, these progenitor cells, called myoblasts, gradually began to develop into muscle cells. Over the next few years, other groups put MyoD to the test and successfully converted a cornucopia of cells—from fat and liver cells to brain and bone tumor cells—into myoblasts.

Studying the paper nearly a decade later, it was all Yamanaka needed to justify his decision to pursue reprogramming. Overexpressing that gene in fibroblasts, he reasoned, it should be possible to convert fibroblasts into muscle. "So from those precedent examples," he says, "I thought we might be able to induce embryonic stem–like cells from somatic cells by introducing one or a few transcription factors."

Yamanaka never deluded himself into thinking he would be able to understand how such a reversal would work—he only needed to be convinced that biologically it could. "Even without knowing the detailed mechanism, it is very surprising to know that only one gene can convert an antenna into a leg," he explains. "That was good enough for me to believe that a similar thing can happen in making ES-like cells."

But first he had some rather mundane laboratory obligations to fulfill. As a condition of his new post at Nara, he was charged with establishing and maintaining a population of genetically modified mice. The institute's administrators clearly saw more value in Yamanaka's mouse managing skills than in his radical ideas about reversing development in cells. As Gail Martin had predicted in the early 1980s, the most practical application of her breakthrough in establishing mouse ES lines was their utility in generating knockout mice. And that's exactly what every scientist at Nara was hoping Yamanaka would do for them.

Given his experience at Gladstone with breeding mice that had specific genes kicked out of their genomes, he was now expected to serve as the institute's bespoke mouse man, supplying its scientists with whatever variety of custom-tailored genetically modified mice they required.

It wasn't the most glamorous job, and it certainly chewed up many of Yamanaka's early days at the lab. But like the time spent at Gladstone studying NAT1, it would prove in retrospect to be more of a boon than a curse.

"It was kind of good for me," he can say now, his smile revealing that at the time he was far from feeling the same way.

In order to get adept at filling out his colleagues' orders, Yamanaka wanted to have as much experience generating the mice as he could, hoping that, as the old adage goes, practice would make perfect. He decided to use his own laboratory project as a trial run, making knockout mice that he would need anyway as he began to comb through the enormous library of genes that might be involved in reprogramming mature cells. His plan was to delete each of the candidate genes he identified as being important in reprogramming and then generate mice from them to observe whether their development proceeded normally.

To find those genes, Yamanaka was betting that the same factors keeping mouse embryonic stem cells in their undifferentiated, embryo-like state would also work in forcing differentiated cells back to a similar condition. "We thought many of these important factors should be expressed specifically in mouse embryonic stem cells," he explains. "So that's where we started." Culling libraries of genes and proteins made by mouse ES cells, the team identified a gene that seemed promising—Fbx15.

So for their first attempt at generating a knockout mouse at Nara, the group decided to splice out Fbx15, essentially killing two birds with one protein. They would gain experience in creating a knockout strain for the Nara core facility, and as a bonus they would get some potentially valuable insight into what effect Fbx15 might exert in arresting cells in their embryonic state.

"The experiment went very well," says Yamanaka, still clearly excited by the achievement. "We got the first knockout mouse line in Nara, and for me, in Japan."

It was certainly good news for Nara's other scientists, since Yamanaka's group had now proven they could create the genetically modified mouse strains, and the researchers could look forward to a steady supply of whatever mice they wanted to study.

Yamanaka himself had less to celebrate, however, since the mice appeared to be perfectly normal. He had thought Fbx15 was a critical factor in maintaining mouse embryonic cells—that if he removed the gene, the mice would not be able to develop. "We thought the embryo should [not survive]," he says. "But the mice were just fine."

Still loath to accept defeat, Yamanaka put Fbx15 to another test, taking ES cells and then wiping out the Fbx15 gene from the cells' genome. If Fbx15 played a critical role in keeping embryonic stem cells in their embryonic state, then certainly splicing it out of their genome would push them out of their undifferentiated state.

"Again, we didn't see anything. The ES cells were just fine without this particular gene," says Yamanaka. "We were naturally very disappointed."

"How long did all of this take?" I ask.

"Like, three years," he says.

★ ★ ★

When Yamanaka was given the Lasker Award in September 2009 for his work in developing iPS cells, he was asked to comment on his journey from failed surgeon to pioneering stem cell scientist. The prize is generally awarded several decades following a scientist's major breakthrough, when it is finally possible to assess, with the advantage of a few years' perspective, the impact a particular contribution has had in medicine. Michael Brown and Joseph Goldstein were recognized in 1985 for their work in the previous decade describing the molecular workings of cholesterol. Dr. Etienne-Emile Baulieu was honored nearly ten years after he developed RU-486, the abortion pill. And Dr. Harold Varmus received his award (and a Nobel) after nearly a quarter-century of work on identifying cancer-causing oncogenes.

Perhaps acutely aware of the relative rapidity with which he achieved success in reprogramming somatic cells, in his essay Yamanaka likened his scientific career to an *ekiden*, the Japanese word for a relay marathon embarked upon not by a single runner but by multiple athletes. From the beginning, Yamanaka was astute enough to realize that attempting to achieve something as radical as reprogramming a somatic cell could not be a solitary endeavor.

So he enlisted the help of an eager graduate student, Kazutoshi Takahashi, for an equally critical experiment. A chemical engineering candidate from nearby Doshisha University, in Kyoto, Takahashi was beginning to show more interest in biology than in his chosen major, but was still doubtful about his ability to refocus his career at such a late stage. "My background was different. I was very afraid of if I can succeed; very afraid," he says in passable English.

When Takahashi began shopping around for a place to complete his Ph.D., Yamanaka appealed on two levels. "It was difficult for me to understand most of the professors' talks. Only Yamanaka's talk was very easy for me," he says. Yamanaka was also a new professor at Nara, and was just building up his lab. "The Yamanaka lab had no history, so I could be his first student. This point was very important for me."

Because Yamanaka intended to generate embryonic stem cell–like cells from already mature cells, he was still attempting to identify, as completely as possible, all of the factors that played a role in keeping mouse embryonic stem cells in their undifferentiated state.

Fresh off his failed experiment with Fbx15, he was casting about for a new way to test his theories. How would he tease apart the factors that were truly essential for nurturing embryonic stem cells from those that were necessary but not critical for keeping the primitive cells happy? And how would he figure out which combinations of potentially dozens of contributing factors were the right ones? His reprogramming *ekiden*, it seemed, had only just begun.

Expanding the initial work that had yielded Fbx15, Yamanaka and

Takahashi booted up their computers and began a massive search of genetic databases from around the world. They were looking for a master list of genes coding for factors essential for maintaining mouse cells in their embryonic state. "We found very, very powerful those in silico analyses of existing databases," Yamanaka tells me, using the term scientists reserve for such database-driven experiments. Then he grins. "And it was very cheap so very good for us, as a new laboratory."

The frugal strategy worked, but it was nothing if not tedious. Their primary resource was cDNA libraries containing snippets of cloned DNA, known as complementary DNA. The genetic fragments are so named because they are made from reading messenger RNA (mRNA) found in the nucleus of cells; mRNA contains only the genes that are active, or expressed by a cell. So Takahashi's main task was to comb these databases for any and all cDNA that were expressed by embryonic cells.

The first list contained one hundred genes.

When Yamanaka described his new project to colleagues, he says, "I often heard, 'Good luck.' But you know, I was not so stupid. I thought the initial process was to identify factors that play important roles in ESC. I thought the final goal was very difficult, and we had a very, very long way to go. But the initial process could be productive, so I could publish at least a few papers every year, and I could survive by studying how ESC maintain pluripotency."

Yamanaka settled himself in for a long and certainly arduous *ekiden*.

★ ★ ★

In another corner of the lab, the student who had generated the Fbx15 knockout mice was still mourning the failure of her experiment. Ordinarily, if knockout mice such as those didn't show any outward signs of their internal genetic makeup, they were sacrificed; without an easy way to detect whatever genetic changes they possessed, it would be a waste of precious resources and valuable time to keep such animals alive. In this case, the mice had written their own death certificates by simply surviving, since the deletion of Fbx15 was supposed to be a lethal genetic alteration.

But, says Yamanaka, displaying a sentimentality that is surprising for a scientist, he could not simply shut the colonies down. "Because those mice were our very first line [of knockout mice], it was kind of difficult for us to simply close—to kill all the mice. So I thought very hard for other ways to utilize those mice."

And in one of those serendipitous marriages of necessity and destiny, he found one.

While Fbx15 had not proven to be the essential reprogramming factor that they had hoped it would be, it was still clearly one of the many ingredients important to creating just the right environment for embryonic cells to survive.

So Yamanaka carved out the perfect role for his precious Fbx15 mice: They would become the living laboratory for testing the embryonic factors that he and Takahashi were amassing. It would be the ideal way to knit together the two tracks of research occurring in his lab, and would make efficient use of what he had initially thought was a strain of mice gone bust. He needed a quick way to test the factors he and Takahashi were coming up with. And the only way to do that was to insert the gene for each of the factors, one by one, into a mouse cell, and then wait to see if that cell would become pluripotent, self-renew, and for all intents and purposes begin to act like an embryonic stem cell again.

But he needed an efficient way to tell whether this reverse development had actually taken place. And that's where Fbx15 proved invaluable. He decided to insert a gene coding for resistance to neomycin—an antibiotic—into the same locus where the Fbx15 gene sat, knocking in instead of knocking out a gene this time. Because Fbx15 was important for embryonic stem cells, Yamanaka figured that if any of his factors reverted a cell back to its embryonic state, that cell would have to begin expressing Fbx15 again. And by yoking that gene's activity to one for neomycin resistance, he could easily pick out the cells that had been reprogrammed, simply by exposing them to the antibiotic; only the stem cells would be able to survive. It was ingenuity and serendipity at work—without the Fbx15 mice, Yamanaka likely would have had to create a strain with similar properties from scratch anyway.

The computational approach was ahead of its time. By designating specific parameters for the genes they were looking for—ones that were turned on only in embryonic cells, but not, for example, in any other cell type—Yamanaka compiled a list of twenty-four factors that seemed to be critical to keeping stem cells undifferentiated. "When he thought of doing it this way, he was way ahead of everyone else," says Eggan, still awed by the strategy Yamanaka used.

Yamanaka knew what had to be done next: Someone had to test each of the twenty-four factors using the Fbx15-neomycin assay. That meant taking the Fbx15 mice (the lab was now crawling with more than a hundred of them, thanks to Yamanaka's soft spot for the animals), breeding them to obtain embryos, and growing a bunch of their skin cells. They could then use retroviruses to slip each of the genes they wanted to study into these skin cells and watch what happened.

"He did something that it would be difficult to persuade a graduate student or postdoc in the U.S. to do," says Dr. Thomas Jessell, of Columbia. "And that is to mix and match those genes and put them in fibroblast cells in the hope of reprogramming those fibroblasts into embryonic stem cells. If I had tried to persuade a graduate student, let's just say it would have met with opposition. Because the chances of that working effectively in a graduate student's career span is pretty low."

Once again, it was a bit of luck that made the work possible. In 2000, when Yamanaka launched the experiment, Takahashi was eager to join the lab, but as a chemical engineer he was still a bit unsure about his skills as a molecular biologist. The task of testing the twenty-four compounds, while tedious to anyone else, was a welcome opportunity for the engineer.

When I ask Takahashi, whom his colleagues call Kazu, how he responded when Yamanaka presented him with the project, he answers quickly, and perhaps with unintended humor. "Thank you very much," he says, was his quick reply. "I wanted an exciting project so when Dr. Yamanaka ask me about reprogramming, I only say yes."

He admits that Yamanaka made sure he was fully aware this could be a career-filling project, one that might not even yield results before Takahashi was ready to retire. But at least he had job security. "Dr. Yamanaka said maybe it will [take] more than ten or twenty years," Takahashi recalls. "But he promise me if I spend more than twenty years, then he does not fire me."

By this time, Yamanaka was ready to move again, now to Kyoto University. He brought with him the Fbx15 mice and, perhaps most important, Takahashi and many of his original lab members. They also had their list of twenty-four initial factors.

"In the beginning, we thought we had to screen a much larger number of genes," says Yamanaka. "Kazu and another lab member of mine were preparing so-called cDNA libraries from mouse ESC. Those libraries contained, at least in theory, all the genes expressed in mouse ESC. So we thought twenty-four factors were not sufficient. But before actually testing those large cDNA libraries, we thought, 'Hey, why don't we test this small library containing twenty-four factors?' We really thought it was just a kind of practice."

That run-through morphed into an unexpected opening night for the lab's small team. "Surprisingly, it worked," says Yamanaka with his typical understated aplomb. "When we put the mixture of twenty-four factors into retroviruses, we saw some colonies that were resistant to neomycin."

"It was very surprising," says Takahashi. "We thought maybe it will require more than a million experiments to find the [right] reprogramming factors. But the first experiment was a hit. It's a miracle."

The results told the ecstatic researchers that somewhere in the cocktail of twenty-four factors was the magic formula for turning adult cells back to an embryonic state. That cocktail might require all twenty-four ingredients, or it might be honed down to a core group of minimum factors to accomplish the feat. Working from that list, Takahashi came up with four factors absolutely required for the skin cells to lose their differentiated state and become pluripotent again—just four.

Could it be that simple? Miracles don't happen in science—at least not often. So Yamanaka asked Kazu to perform the experiment again. And again and again. On pins and needles, for six months Yamanaka canceled regular lab meetings while Takahashi confirmed the results. "For first six months I repeated same experiment more than twenty or thirty times," says Takahashi. "Because it was too simple to be true. A possibility was that it's contamination with ES cells. So I repeated it many times."

It was February 2006 when Yamanaka saw the first neomycin-resistant colonies of cells growing—the very first cells that had turned back time and been reprogrammed from a specialized state to become embryonic again. But his timing was awful. Across the Sea of Japan, in Seoul, Hwang had just been exposed as a fraud; his history-making cloned stem cells from both healthy and diseased patients simply didn't exist. The field was particularly sensitive about spectacular claims, and Yamanaka was painfully aware that his would likely be greeted with more skepticism than awe.

"We immediately knew that we were going to have a hard time publishing this paper," he says. "Because many people should be skeptical about the result; it's too simple. We were kind of afraid of publishing our first paper."

Yamanaka realized that the technique was just too straightforward—unlike the nuclear transfer method that had produced Dolly and that Hwang had claimed to have used to generate his human embryonic stem cells, the reprogramming method was ridiculously cookie-cutter. "All you have to have is a cell culture system and basic technology in molecular biology and gene transfer," says Yamanaka. "Any regular lab in a scientific institute can do this."

It was so easy, in fact, that Yamanaka was reluctant to publish the mouse data until he had accomplished the same feat with human cells, fearful that once his methodology was out in the open, someone else would beat them to achieving the more coveted milestone of converting human cells back to an embryonic state. Reprogramming human cells would kick open the door to an entirely new generation of stem cell–based treatments for disease, and certainly net the discoverer a permanent place in medical history.

As an academic scientist, however, it made no sense to hold back the announcement of such an explosive finding. He coyly leaked out versions of his results at two key meetings—one of the Keystone Symposia on Molecular and Cellular Biology, and the 2006 ISSCR annual conference. Sitting in the audience of the Keystone gathering during Yamanaka's presentation, Eggan was floored by what he was hearing. Having just received approval to clone human cells to make stem cells, Eggan, like the rest of the field, had been convinced that nuclear transfer was the most direct and accessible way to generate stem cells from patients.

Then Yamanaka got up to speak. "We were feeling good about ourselves, and then whammo!" says Eggan. "When Shinya got up there and started talking about this, a lot of people didn't believe it. But I just felt in my heart. . . . I knew it was true. Because basically there was too much data. He hadn't totally mastered it, but he had it."

By the time his results were published later that summer in *Cell*, Yamanaka had introduced a new type of stem cell to the world, and coined a new name that will dominate scientific and clinical journals for decades to come. Wanting to avoid at all costs any association with the politically and ethically loaded embryo, he named his cells induced pluripotent stem cells, or iPSC. "I thought if we had a name which is similar to embryonic stem cells then people in the government may think they are too similar to embryos, so they may decide for some regulation on these new stem cells," he says. "So I decided no *e*."

★ ★ ★

For stem cell scientists, it was as if the coal-powered train they had been riding had gone supersonic. Instead of enjoying a leisurely journey toward understanding development, with scheduled stops along the way to appreciate and absorb their newly encountered discoveries, they had whizzed past all the familiar mile markers and been deposited, disoriented and heads still spinning, at their spectacular if still unbelievable destination.

For researchers like Eggan and Thomson, Yamanaka's stunning achievement was like a scientific taunt. Would it work with human cells? Who would be the first to find out?

In Kyoto, Yamanaka was already pushing his team to translate the work from mouse cells to human skin cells, using cells obtained from a tissue bank. Thomson, unbeknownst to the rest of the world, had been quietly following up his own stem cell work with a reprogramming experiment of his own. For six years, his lab had been pursuing the same brew of factors that would make stem cells out of human adult cells. Yamanaka's presentation at that Keystone meeting, he tells me, "wasn't a wow moment. It was a scrape-yourself-off-the-floor-and-hope-you-have-something-to-salvage kind of moment." Because Yamanaka did not reveal all four of his secret ingredients in that talk, Thomson had no way of knowing whether he and the Japanese team had stumbled upon the same list of factors.

In Boston, Eggan took up the challenge, albeit too late. "I was probably more gung ho than most about the iPS technology," he says. "I tried desperately to get my lab to [copy] the human equivalents of as many different embryonic stem cell genes as we could. It wasn't enough, but I knew we had to get on this technology right away. And I'm glad we did."

In November 2007, Yamanaka and Thomson simultaneously announced their success in making human iPS cells. The two groups used different formulas for achieving their feat, but had two of the four factors in common. Both, however, had taken the critical step. Now it appeared that all of the theoretical posturing of stem cell advocates might actually come to fruition. Anyone could make stem cells from any adult—healthy or otherwise. No embryos, no politics, no controversy. And with the ease of making iPS cells, researchers were also dreaming up ways they could watch diseases as they occurred in a dish, in a way that had never been possible before. Within a year, Eggan would become the first to put this idea to the test, conjuring up motor neurons, the nerve cells that deteriorate in ALS, from the skin cells of two elderly patients. "For all of us," he says, "it's a new day."

12

Looking Ahead

Douglas Melton is in full professor mode—chalk in hand, animated—and it doesn't seem to matter that he has a class of only one. As he stands in front of the small chalkboard in his office (installed, it seems, for just such tutorials), his words flow in a swift torrent of ideas and thought experiments and scientific explanations.

He is mapping out for me the huge shift in focus that's occurring in his lab; as these things go, it's seismic. It's a clear fall day, and after overseeing the weekly lab meeting, Melton has called me in for a little science lesson. This morning's presentation, by Sinisa Hrvatin, one of his Ph.D. students, focused on the mechanics of growing up insulin-producing beta cells. As much as Melton's name has become synonymous with stem cell research, more and more, the newest members of his lab aren't working with the cells per se, but on the enigma of what makes beta cells happy—which milieu of growth factors, nutrients, and other X factors keep the cells healthy and able to reproduce and do their job of secreting insulin. It's a reflection of how quickly the science is evolving—in the past few years, Melton has been folding in more doctoral candidates with diverse interests, which at first blush might seem far afield of either stem cell science or diabetes.

But, he assures me, there is a direction to his seeming distraction, and it emerges almost directly from Yamanaka's breakthrough. Milestones like that are similar to a hole in a dam—you don't realize how much pent-up force is building until a tiny opening allows some of the energy to trickle through. In a testament to the robustness of the science that was simply waiting to burst forth, once Yamanaka proved that human cells could be reprogrammed and given a chance at a second life, the advances and ideas inspired by iPS technology generated a torrent of creativity and innovation.

With iPS cells, the barriers, both scientific and political, that had stood for eight years in front of researchers like Melton began to melt away. Because it uses adult cells and never needs an egg or an embryo in order to grow, the technology easily sidesteps the ethical hurdles that had kept embryonic stem cells in a holding pattern for so long. In fact, the most vocal anti-embryonic research representatives even attempted to spin iPS technology as a validation of their argument in favor of alternatives to destroying embryos. The net result, however, was a boon for stem cell scientists: Even while continuing to restrict funding of human embryonic stem cells to those approved by Bush in 2001, the NIH, after 2007, allowed grants for studies on human iPS cells.

And scientifically, because the technique required only the most basic of gene manipulation skills, iPS became the everyman version of stem cells. Unlike the case with embryonic type, which require some experience with handling eggs and embryos and some facility with the micromanipulator—and interminable patience—practically any molecular biology student can mix up a batch of iPS cells from a skin sample. Such an easy entrée was critical to infusing new ideas and new energy into the stagnating field. "I'd like to think that young people think it's just plain cool that my skin cells can be turned back and can make a whole other copy of me, potentially," Melton says.

The very limitations of embryonic stem cells—the scarcity of human eggs to generate them, the ethical and moral challenges in creating the embryos that were the only sources for the cells—are the strengths of the iPS technology.

And for Melton, that technology meant that finally—finally—he would have a chance of confronting the foe that had shadowed his family for so long. With iPS cells, he could realize his goal of watching diabetes unfold before his eyes in a petri dish. It has been, Melton admits, a long journey, and one with its share of detours, but ultimately, he hopes, worth the wait.

★ ★ ★

Almost immediately after hearing Yamanaka's presentations at the Keystone and ISSCR meetings, Melton's mind raced to map out the possibilities. He had made slow but significant progress in coaxing ES cells along the stepwise progression from stem cell to endoderm to pancreatic progenitor to exocrine cell and eventually to an insulin-producing beta cell. With that approach, he was hoping to first establish how a normal insulin-producing cell arises, and then, perhaps using nuclear transfer, generate beta cells from diabetic patients and compare them. Only then could he pick out the genes that were acting differently among the two groups and perhaps learn something about how to turn the disease around.

With Hwang's failure, however, there is still no way to use nuclear transfer to make such patient-specific stem cells. But the iPS technology gives him the ability to leapfrog directly to the culprit cells, so he can now take a skin sample from a diabetic patient and nurture it in the lab until it becomes a beta cell, all the while carefully monitoring its development and exposing any aberrations or deviations along the way that would make good targets for a drug. It's the disease-in-the-dish model that first lured Melton to stem cells. In fact, it is this feature that scientists see as being the first real and practical benefit of the technology—not transplants of tissues made from stem cells, which became the focus of the political debate but are perhaps at least a decade, if not more, away from being routinely tested in people. Especially with diseases such as type 1 diabetes, Parkinson's, and Alzheimer's, the iPS cells provide an ideal diorama for watching the pathology unfold because there likely isn't just one gene responsible for triggering the disease pathway, and genes are almost certainly interacting in complete ways with environmental factors (diet, stress, and the like).

The scientific community didn't wait long to put its new tool to use. Handcuffed too long by political and regulatory restraints, they pounced on the chance to finally make stem cells free of regulatory and political baggage. Within months of Yamanaka's publication, Eggan's team cultivated the first cells from patients using the iPS technique. Taking skin cells from two patients suffering from ALS, Eggan successfully doused them with the four Yamanaka factors and reprogrammed them back to a pluripotent, embryonic state. Then he prodded these cells to develop into motor neurons, the nerve cells that gradually lose function in ALS. The same year, Melton's team did the same for diabetes, creating the first stem cell lines from two young boys with type 1 diabetes.

Yamanaka's breakthrough was a defining moment for stem cell research, a powerful proof of principle of an entirely new way of studying and teasing out the causes of disease. "If we can observe disease processes, then we have not just opened the door but the garage door in understanding the cause of why that cell screws up, what goes wrong— and we can do drug screening and study the gene expression of these cells," explains Melton. "We can't do any of those things in a person. We can't take a Parkinson's patient and go into his brain. And even if we could, we'd have to go in every ten minutes from the time they were born to when they were sixty-five and developed the first symptoms, because we wouldn't know when the problem occurred."

As proof of the power of this idea, at the eighth annual meeting of the ISSCR, the sessions overflowed with 230 reports from labs around the country that had created their own dioramas of disease. These included those for long QT syndrome, a disorder affecting the electrical signals in the heart; familial dysautonomia, a congenital disease affecting nerve development; LEOPARD syndrome, a genetic abnormality affecting a constellation of organs, including the skin, heart, face, and genitals; spinal muscular atrophy; fragile X; and dyskeratosis congenita, a rare bone marrow disorder that results in premature aging.

No one is arguing that iPS cells are ready for the clinic; nor are they being primed to replace damaged heart cells or replenish dwindling beta cells anytime soon. They are still far too much of an unknown quantity for that. Researchers are still trying to figure out the answers to critical questions about the reprogramming process—about how faithfully its embryonic state mirrors that of true embryonic stem cells, and whether iPS cells can indeed generate every one of the more than two hundred different cell types in the body. But they have already made a crucial first step toward that goal.

Yamanaka relied on viruses and genes to reprogram his cells, but even he realized that while the technique was fine for making iPS cells for research purposes, it was too unsafe for generating transplant-ready tissues. The viruses' tendency to randomly integrate anywhere in the cell's genome make them molecular hazards, since they can interrupt key metabolic and development processes, and the genes themselves are known to promote cancer. Groups around the world have been racing to find replacements for the Yamanaka factors, looking for less pathogenic vehicles for the genes, for example, or even seeking compounds that would substitute the entire virus-gene package with molecules with a better safety profile. Yamanaka himself managed to swap out the viruses for a more inert string of DNA, while Melton replaced two of the four reprogramming genes with safer molecules.

But it wasn't until late 2010 that a truly viable alternative emerged. And like so many brainstorms in science, it's one that seems so obvious in retrospect.

The problem, essentially, is a mechanical one: how to saturate the cells with enough of the Yamanaka factors to push them to reprogram themselves. Genes are too unstable to do the job, but the DNA in genes is processed by cells into another genetically coded form known as RNA. RNA is not part of a cell's genome, but floats around inside a cell until it binds to specific structures that turn the RNA into a final protein product. Stem cell scientists were scrambling to find small molecules to replace these proteins, or safer ways of integrating the genes that coded for them—but couldn't simply dousing the cells with the proper RNA versions of Yamanaka's factors accomplish the same thing?

Surprisingly, not too many researchers had hit upon this approach. But Derrick Rossi, an immunologist at Children's Hospital Boston and a part of the Harvard Stem Cell Institute network, thought it sounded like a promising strategy to pursue. "Amazingly, with all the approaches people were trying [to reprogram iPS cells], we never saw anybody propose it in the literature," he says. He soon found out why. It turns out cells have a natural defensive mechanism against having large amounts it deposited within their confines; viruses, after all, are packets of tightly woven RNA strands. Within hours of injecting the RNA for Yamanaka's factors into some mouse cells, Rossi was disappointed to see it chewed up by interferon, an antiviral agent that the cells began churning out in response.

Undeterred, Rossi thought he had a work-around. By attaching a protective string of chemicals to the RNA fragments, he could give them a molecular disguise, making them invisible to the cell's immune sentries, thus sparing them from being tagged for destruction. To his surprise, the strategy not only worked, but it worked significantly better than Yamanaka's original method of turning mouse skin cells into iPS cells. In fact, the RNA strategy boosted the efficiency of the process by up to a hundredfold.

With a potentially safe and robust way of making stem cells for human testing, for Melton the possibilities are becoming deliciously tangible. As the father of two diabetic children, he can almost taste the changes that are coming as iPS technology matures and exposes ever more about how diabetics lose their ability to make insulin. The breakthroughs, in fact, were enough of an inspiration to trigger another about-face for his lab, the subject of the tutorial he's giving me in his office.

★ ★ ★

Melton's grand plan has been building for more than a year, and represents only a slightly less radical version of the turnaround that he made in his career seventeen years ago. As always, it has everything to do with chasing that still-elusive cure.

What it boils down to is this: After establishing himself and his lab as one of the epicenters of stem cell research, Melton is now turning his attention toward solving what he sees as the other half of the problem in type 1 diabetes—autoimmunity.

"There are two problems with type one diabetes," he explains, diagramming the challenges quickly on the board. "The first is, no beta cells. The second is an autoimmune attack. If you cured either one of those, it would be insufficient to cure the patient. If you blocked the immune attack, since they don't have any beta cells, they still can't make insulin. If you provided beta cells, they are still going to be killed by the immune attack."

He pauses, leaning on the bookcase on the wall adjoining the board. "Among the mistakes I made in the first five to ten years working on the problem," he admits, "is that I for complicated reasons didn't work on autoimmunity. One, I didn't really understand it much; two, I thought that making beta cells itself was a monumental problem; three, I assumed that many immunologists were fully focused on it and that the problem would be solved by now. All of those things turned out to be wrong."

For nearly two decades, Melton has relentlessly focused, perhaps even become obsessed, with addressing only the lack of beta cells. Initially, that meant searching for the beta cell reservoir, the mother cell pool of pluripotent cells somewhere in the pancreas that gives rise to the body's supply of insulin-making cells—the adult stem cell of beta cells.

Most scientists at the time were convinced that every tissue type in the body had some sort of stem cell provenance—after all, the entire immune system originated from blood stem cells that could regenerate every white blood cell, macrophage, lymphocyte, T cell, and B cell that the body needs to fight infection. Millions of leukemia patients can attest to that, as bone marrow transplants have held off blossoming tumors of the immune cells by basically wiping out the diseased system and starting all over, nurturing the few healthy stem cells from the bone marrow and reinfusing them into the patient until the whole army of immune cells has been built back up, cell by cell.

"Everyone just assumed that there is an adult stem cell for the beta cell," he says. But Melton didn't have high expectations—if such a cell did exist, it wouldn't be easy to find. It isn't as if the pancreas is like the skin or the bone marrow, both of whose cells are shed and reborn at an astonishing rate to replenish constantly dying ones. No, the pancreas is a bit more sentimental, holding on to its cells for as long as possible, and giving them up only when they have outlived their usefulness or are too spent to continue on. If the pancreas did have such a reservoir, it wouldn't be very active, since pancreas cells don't turn over as much.

But after searching for nearly a decade, as experiment after experiment failed to isolate the beta stem cell, Melton finally made a stunning discovery: The beta cell has no adult stem cell. He found instead that the entire population of beta cells in the pancreas comes from preexisting beta cells that continue to divide to produce more progeny. It wasn't great news for diabetics, since their cells were systematically getting killed off by the immune system faster than they could replenish themselves.

"This changed everything," he says. "This said that if a patient lost their beta cells, and there is no stem cell, then they are in big trouble."

Beta cells, then, are a valued commodity, the body's equivalent of precious molecular metal since the entire beta cell population also derives from the embryo. But there is growing evidence that even in type 1 diabetics, whose population of beta cells presumably hovers near empty, a residual population of ragtag cells may be valiantly holding forth and churning out insulin that, alas, will never be enough to meet the crushing need represented by the perpetual fluctuation in glucose levels. In fact, in autopsies of diabetic patients who have died from other causes, such as car accidents, many still retain a legacy of beta cells. "When you are little, you got a dollop of beta cells and those beta cells are the source of all new beta cells from now until you die," Melton says. And while confirming this might seem like a purely academic question, he emphasizes that it's not, because how scientists frame their questions determines the direction of research into diseases such as diabetes and, ultimately, how quickly they get to a cure. Take, for instance, the NIH, the largest funder of scientific research in the United States. "The NIH has spent tens of millions of dollars—I daresay hundreds of millions of dollars—looking for the adult stem cell for the beta cell," he says. "But it's never been found. But because they were doing that, nobody studied what I now say is the key thing, which is what makes beta cells divide."

If there is no beta cell stem cell, then the question of how that initial allotment of beta cells maintains a lifetime's worth of insulin-making cells becomes the critical question for understanding type 1 diabetes. Understand that, and you can begin to analyze what goes wrong in these patients.

All of this is bringing Melton around to how he is trying to solve the problem of missing beta cells. As he sees it, there are three ways now to make more beta cells—by prompting the existing cells to replicate, by generating new beta cells from stem cells, or by convincing other pancreatic cells to turn into beta cells—none of which, he concedes, is very efficient at the moment.

But Melton proved that expanding that population of preexisting beta cells could in fact cure an animal with what he calls a simple experiment with a very complicated genetically engineered mouse. Melton took his first steps toward his holy grail—watching type 1 diabetes unfold in the human. The mouse, a marvel of modern genetics that can selectively kill off its beta cells, allows him to ask and hopefully answer critical questions such as: How many beta cells have to be destroyed before the disease sets in? Is there a minimal population of beta cells needed to avoid the disease, and is this threshold the same in every patient? And if so, would reaching this threshold be sufficient to treat or cure someone?

"It's quite a clever design," he says, back at the board to sketch out the animal's intricate genetic makeup. Outwardly, the mouse looks and acts like a normal animal; it's not diabetic and can properly control its blood sugar. Spike its water with the antibiotic doxycycline, however, and a remarkable transformation takes place in its pancreas. That's because it contains a little snippet of DNA that links its insulin gene to a portion of the genome that is responsible for activating tetracycline production, which indirectly leads to beta cell death. If the mouse gulps down doxycycline from its drinking water, the drug binds to the tetracycline activator and turns on production of diphtheria toxin A, which kills the beta cells. More doxycycline in the water leads to more complete destruction of the mouse's beta cell population—essentially leading to a way to create mice that are more, or less, diabetic.

After inducing diabetes in a series of mice this way, Melton then stopped feeding them doxycycline and watched what happened. To his surprise—and delight—the mice eventually stopped being diabetic and went back to metabolizing sugar normally. "If you killed all the beta cells, the mouse's blood sugar stays high and it dies," he says. "But if you only killed eighty percent of the beta cells, now you have a residue of twenty percent—and this is really cool—" he says, getting excited, "the blood sugar level over a period of six weeks goes back down to normal. The animal cures itself. And we guess the way that it does that is that it boosts replication of those residual beta cells."

So from Melton's perspective, this residual population is akin to a beta cell seed stock—if only he could get at it. If these few cells can be amplified, then perhaps at least some of the burden of type 1 diabetes can be lifted. If more of the patient's own beta cells can take over their normal function of reading and responding to glucose, then maybe, just maybe, these patients can reduce their reliance on insulin injections.

Melton has assigned the task of understanding this process to his own Ponce de León, Justin Annes, who holds both a medical and a Ph.D. degree. That allows him to see both sides of the disease—the patients affected and the molecular processes that make them sick. As an internal medicine specialist at nearby Brigham and Women's Hospital in Boston, he primarily sees type 2 diabetics, but he has an interest in understanding some of the basic mechanisms that might tie type 1 and type 2 diabetes together. Annes is now focused on analyzing this beta cell mass. "If you look at diabetics or even people with impaired fasting glucose postmortem, they have a reduced beta cell mass," he says. "If you take siblings of type 2 diabetics who are nonobese and look relatively healthy, challenge them with glucose, and look at how much insulin they produce, they also tend to have reduced beta cell mass, which suggests an intrinsic, probably hereditary deficit related to how many beta cells you have. So Doug and I are ultimately both interested in how beta cell mass is increased. We share a common goal; we just look at it from two different perspectives. I'm interested in doing it for treating type 2. He wants to understand how to increase beta cell mass. So we might end up transplanting them, and to improve the amount of beta cells we get per donor, we can grow them outside of the body and try to expand them [in culture first]."

The idea, from Melton's perspective, is to better understand the signals that prompt the stalwart few remaining beta cells in type 1 diabetics to divide. One possibility, he says, is that higher glucose levels in the blood could be driving the beta cells to work overtime. As blood sugar rises, the beta cells scramble to keep up. Like a factory struggling to maintain the same production output with half its staff, the beta cells start to fall short. When they can't keep up, they recruit more help by dividing to repopulate the beta cell contingent. In most cases, however, they are too few and too late, and as quickly as they divide, they fall behind the body's need for more and more insulin. That possibility implies a systemic, body-based signal for beta cell replication.

But there is also another, equally plausible driver that has nothing to do with blood sugar at all. "What if the cells are like people, and they want a little community around them?" Melton muses. "After most of the beta cells are killed off, the remaining ones look around at their neighbors and say, 'Hey, where is everybody? We were a nice big juicy islet, and now there are only twenty of us left, so let's reproduce.'" In this case, the signal isn't a metabolic one, but a physical one. And it isn't systemic, but very local.

But which one is right? And why is it important? Essentially, it matters because the answer dictates which treatment strategy makes the most sense. If the cells are responding to a body-wide signal to divide frantically, then simply dumping more beta cells into the pancreas won't necessarily solve the problem. It's the signal that regulates the beta cells' activity that should be the target.

But if, on the other hand, the beta cells can be coaxed into dividing simply by the presence of additional beta cell neighbors, then the prospect of infusing a new population of these cells into the organ makes sense.

And getting at the answer is just the type of puzzle that Melton relishes. Still at the chalkboard, he outlines another very simple experiment with the same complicated mouse. This time, he grafts a healthy mouse's islet cells, which are home to the beta cells, into the kidney capsule of the genetically modified diabetic mouse. This procedure immediately brings the mouse's blood sugar levels down, as the donated islet cells take over for the beleaguered few that remain in the original animal. If the signal for beta cells to replicate is systemic and governed by blood sugar levels, then those remaining beta cells in the diabetic mouse would not divide further, since the glucose levels are now stable and there is no alarm. If, on the other hand, the signal is local, then, says Melton, "they are still missing their neighbors," meaning they will still sense the dearth of beta cells, and they will start to divide to replenish themselves.

As that study continues, Melton isn't putting aside the basic question of how to generate beta cells in the first place. Running on a parallel track to this work are the studies that have garnered him the most attention from the scientific community and the public alike over the past decade—generating a rich supply of the insulin-producing cells, first from embryonic stem cells and now from iPS cells, and even more recently, from other pancreatic cells.

If the embryo is the source of a person's entire lifetime supply of beta cells, then it's understandable why Melton has devoted nearly twenty years to studying ways to coax the insulin-producing cells from early development—and why he has become an outspoken advocate of human embryonic stem cell (hESC) research. Very early in the process, he made the crucial decision to work with only human cells, bypassing the traditional step of testing and validating his theories and experiments in mouse models first. "If I have any mistaken prejudices, then these are an unrelenting focus on human cells here," he admits. "Looking back, some might say I should have spent the last decade not messing around with the politics of hESC and just done this in the mouse. It's a prejudice—in all the good and bad senses of the word—that we want to cure humans. I am not convinced that if we did this in mice it would translate immediately to humans. So we work on humans every time we can."

Which is how the "balloon" diagram was born. Since Melton debuted it soon after his son's diagnosis, as a way to help himself and his lab members to visualize the stepwise process of re-creating a human beta cell, it has appeared at nearly every update Melton gives at scientific conferences. It details a five-step process that moves from a human embryonic stem cell to a beta cell, with each step depicted as a colored circular cell. When he first used it in his talks, Melton had only theories about how to move from cell to cell. First on the list was pushing the embryonic stem cell to become endoderm, one of the three germ layers that eventually give rise to the remaining cells and tissues in the body. Endoderm then needs to be directed into the pancreatic progenitor, or the pancreatic cell that sires not just the insulin-producing beta cells but the other cells in the organ as well, such as exocrine cells that make digestive enzymes. This progenitor then develops into subtypes of cells, with one lineage destined to create the specialized beta cells that respond to glucose and secrete insulin.

After nearly a decade, Melton has finally teased out two of the five steps, which, he insists, isn't as bad as it sounds. "This idea has not been achieved as quickly as I thought I would," he admits. "But everything we have learned tells me this is possible, [and] I think we have broken the back on that problem. I think it's going to happen; it might still take ten years, but it wouldn't surprise me if we did it within a year. That's a solvable problem."

As Melton sees it, now it's a technical problem, a matter of figuring out, through careful screening, which chemicals and which compounds are necessary to direct each step of the differentiation. Solving that puzzle now lies in the hands of Shuibing Chen and Sinisa Hrvatin, who are, slowly but surely, exposing the chemical steps necessary to complete the progression from stem cell to beta cell.

Yamanaka's iPS cell breakthrough, however, made the painstaking steps symbolized in the balloon diagram seem passé. With the four Yamanaka factors, Melton suddenly had a relatively easy way of scooping up all of that complex interplay between one gene and another, and between genes and other factors that he was trying to capture in just the right combination, in one fell swoop. All he had to do was create a stem cell line from someone with diabetes. Then there would be no need to guess whether diabetes would develop—it simply would.

Using what he had learned from his work to date in isolating the factors that drove beta cell formation, Melton applied those same factors to a few skin cells from two diabetic boys. Just months later, the dishes contained iPS cells. And these so-called DiPS cells, for diabetes iPS cells, are now the key players in Melton's drama.

But that doesn't mean that he has given up his balloon diagram, or the ES cell route it depicts. Melton is among the majority of researchers who feel that cells made from both ES and iPS cells will be crucial in fully understanding human development and disease—for example, it's still not clear how faithfully the four transcription factors actually reprogram cells via the iPS method. And based on studies of gene activity in the two stem cell lines, it's clear there are differences between the two methods. It's just not obvious yet how important those discrepancies are.

★ ★ ★

Even as he continues his more straightforward efforts to reprogram beta cells, as a testament to just how groundbreaking Yamanaka's reprogramming feat truly was, Melton was inspired to play around with the idea of development with a detour—one that he hopes will become a shortcut—on his beta cell journey. If it were possible to reprogram an adult cell, he thought, why start with an embryonic cell and urge it to jump through the myriad developmental hoops to grow from endoderm to pancreatic progenitor to exocrine cell and finally to an islet cell?

"The excitement of the Yamanaka approach is that we can take cells from a patient and turn them back to become stem cells," says Melton. "But we are still left with the puzzle of how to tell those cells to become a heart cell or a pancreatic beta cell. So we asked why [we should] have to go all the way back to the beginning. Why not take a differentiated cell, like a skin cell, and directly reprogram one cell to become another?" Why, in other words, should a reprogrammed cell have to pass through an embryonic stage in order to begin its second life? Why not jump in halfway, or nearly all the way, with a pancreatic progenitor and coax it along the far fewer steps to becoming a beta cell?

Biologically, it made sense. Yamanaka had proved that it was possible to bring a cell back to its beginnings, and John Gurdon's experiments with frogs had shown that every cell retained its full complement of genes but simply turned off those it no longer used. And indeed, in 2008, Melton answered his own question. Using a set of nine genes involved in making beta cells, he took a type of pancreatic cell normally responsible for churning out digestive enzymes and successfully convinced it to produce insulin instead, just like a beta cell. It's a repurposing Melton likes to call an extreme makeover of cells, and, he says, "maybe we don't need to make stem cells and direct them down all the paths of differentiation.

Maybe we can go directly to reprogramming the cells that we want in the future."

It's certainly a more efficient and ultimately less buggy way of generating new cells. But it shatters yet another of the preexisting notions about development that had become ingrained in biology. "It challenges people's prejudices about how hard it is to reprogram," says Melton, "and about the stability of the differentiated state. But I'm hoping people will rethink their whole strategy and use this approach to make cells that might be missing in a patient."

He needn't worry. Already the idea has caught on, and researchers at Stanford University have taken skin cells and pushed them directly into nerve cells using the Yamanaka factors. At the Gladstone Institute, Deepak Srivastava, a heart researcher, accomplished a similar feat by turning human skin cells directly into heart muscle that could one day replace scarred and damaged tissue following a heart attack. Other cells, Melton hopes, won't be far behind.

Which brings him to the need for the radical shift in focus for his lab. In gathering the players he needs to complete his diabetes diorama, Melton realizes that the beta cell is only one, albeit the primary member of what he now sees as a multiplayer cast. Relatively soon, he realizes, if all goes well, he will have his long-sought-after beta cells. The next question is what he will do with them.

Transplanting them into diabetic patients would seem like a logical move, but it's becoming clear that an influx of new beta cells would still not solve the problem, because if the disease represents both a lack of beta cells and an aggressive autoimmune attack, what would prevent the body from dispatching the new beta cells exactly as it had the original ones—that is, selectively seeking them out and destroying them?

For that answer, Melton is turning north of the pancreas to the thymus, where the body's immune cells, including the killer T cells, are processed and matured before they roam the body in search of targets. So it's in the thymus that some of these cells are somehow given faulty instructions and taught to single out and destroy the body's own beta cells. Melton is planning on using iPS technology to generate these thymic cells from diabetic patients, combine them with immature immune cells that are processed by the thymus, and introduce them, along with a dollop of beta cells from type 1 diabetics, into a mouse. And with his fingers crossed, he's hoping that the mouse will develop diabetes and that he'll be able to see exactly how that happens. "We're going to put all these things into what I like to call a living test tube," he says. "I'm placing a huge bet on a very risky thing; I know that. There are people who think this is simpleminded—they say that immunology is way too complicated, and that I'm not an immunologist. Both of those things are true. There is no reason to believe this would work. But it kind of forgets that often advances in science come from someone who doesn't really know what they are doing and asks a simple question." He pauses, considering. "If it fails, I won't shoot myself, but it will be a huge disappointment."

★ ★ ★

Even if Melton's grand plan proves overly ambitious and the diabetes diorama doesn't play out quite as he expects it to, the effort may not be entirely in vain, because the diabetes iPS cells themselves may unlock a library's worth of new compounds that could alleviate or slow down the progression of the disease.

Indeed, that's what Lee Rubin, a biotech veteran and neurobiologist, is counting on. Ensconced just one floor away from Melton, Rubin is the Harvard Stem Cell Institute's own drug czar. Most of his time is spent in a spacious basement laboratory that would be the envy of many small biotechnology companies. Outfitted with the latest screening robots for testing thousands of compounds at a time, this lab is where the institute's scientists' earliest hunches for treating disease are put to the test.

Energetic and fast-talking, Rubin, a native New Yorker, is Melton's personality opposite. Where Melton exudes calm and unflappability, barely raising his voice even to make a point, Rubin is an expressive bundle of exuberance. It's that energy, though, that is a perfect fit for what Melton wants to do in pushing stem cell technology in as many different directions as possible.

For it is Rubin's work, and experiments like his that are being conducted in similar screening centers at biotech start-ups and pharmaceutical labs around the world, that is likely to yield the first true products of stem cell science that patients will be able to see, use, and benefit from.

It's taken some time, however, for the rest of the world to come around to Rubin's thinking. With the very survival of stem cell research at stake in the early days of the Bush administration, many of the field's researchers found themselves reluctantly pulled into the cell transplant camp. That is, they started talking more about the promise of stem cells to generate new and healthy cells to replace diseased ones, despite the fact that this application had the longest time table and likely wouldn't yield any promising results for a decade or more. "My big worry about the field from the very beginning, when everybody was obsessed with cell replacement therapy, was the 'when' question," Rubin says, speeding through his argument. "My worry was that it wasn't going to be happening quickly, and people, including funding agencies, would get discouraged if it took fifty years to get anything out of the so-called magic power of stem cells.

"So I was thinking about them in an entirely different way—that you could make cells involved in disease from stem cells, and do things with those disease models of cells that you couldn't otherwise do well, or at all."

Such as, he says, throwing them up against a panel of potential drug compounds to see if any of those agents would have an effect on slowing down or even reversing whatever aberrant process was causing the cells to act abnormally. At his Harvard center, Rubin has a library of thirty thousand such compounds, carefully culled to include those involved in the most fundamental and important molecular processes in the body, with which to work.

He is taking, for example, the motor neurons generated from ALS patients that Kevin Eggan's lab created and putting them to the test. What he is looking for is a compound that will keep the iPS-derived motor neurons from dying. And based on work from Eggan's lab, which showed that glial cells might be slowly killing off these neurons in people with ALS, Rubin is looking for agents that can counteract that toxic effect, or protect the motor neurons from succumbing to the destruction.

"All of this suggests a whole new paradigm of discovery, where you basically use stem cells first to understand disease better, and second to test potential therapeutics on the actual cell that's dying and not on cells that are unaffected. This," he says, "is a really big change; this isn't a little thing."

Financially, he argues, it could even save millions of dollars in drug development. To prove the point, across the country at iPierian, a biotech company he cofounded with fellow HSCI members Melton and George Daley, scientists took iPS-derived motor neurons from patients with spinal muscular atrophy and tested them against the fifteen or twenty compounds that have been approved for human testing against the disease. All of these compounds passed rigorous animal testing for safety, but all except one were toxic against the human cells.

That's the kind of information that iPS technology can now make available to both streamline and improve the drug development process, and it's certainly what the pharmaceutical companies are hoping for. With the cost of trials to test new drugs skyrocketing and the pipeline for so-called blockbusters drying up, executives are bracing for a drastic remodeling of the industry, which will almost certainly involve stem cells in some way. Pfizer, for instance, is investing one hundred million dollars over three to five years in their Regenerative Medicine unit. And others, including Merck and GlaxoSmithKline, are joining the trend with their own stem cell–based pipelines for new drugs.

It wasn't as if the big pharma giants were eager to jump on board the stem cell express. After all, scientists like Melton are talking about curing a disease like diabetes, which means essentially cutting off the flow of patients in need of therapies that are the lifeblood of the industry. At the same time, however, the drug makers are fearful of being left behind when stem cell technology starts its inevitable creep into everything from drug discovery to medical therapies. "For regenerative medicine, we and everybody else don't know what the cost is going to be," says Ruth McKernan, of Pfizer, whose labs in Cambridge, Massachusetts, are just a few miles down the Charles River from Harvard. "But if we don't start learning, we're going to have to spend a huge amount of money later, and we may find ourselves not able to contribute to medicines in this space at all. It's parsimonious in the extreme to stand on the sidelines and say, 'We're just going to wait until somebody else does it, and then we're going to buy in.' Because you can make a very, very expensive mistake."

★ ★ ★

Stem cell scientists are all familiar with a rather famous video of rats walking on all fours. It starts out with a rat pulling itself along, seal-like, on its forelegs, dragging its tail and hind limbs. A second clip shows a similar rat that looks a bit more normal, doing what these animals normally do—sniffing, twitching its nose, and walking, if not confidently, then at least with seeming purpose, while inspecting its space.

On its surface, the recording doesn't seem that extraordinary. Except that both of the rats in the video had had their hind legs paralyzed with similar injuries to the spinal cord. The second animal, however, had received an injection of a stem cell solution that managed to restore its ability to walk again on all fours.

Known as the "running rat video," it made the rounds in the heyday of the stem cell debates on Capitol Hill in the early 2000s, and played a critical role in convincing Californians to pass Prop. 71 to fund embryonic stem cell research in their state. In the process, it turned the biologist responsible for it into a media darling.

Eager, articulate, with a boyish charm, Hans Keirstead has more than a passing resemblance to the actor Michael J. Fox. He wears his wavy brown hair slicked back from his face and is quite comfortable and engaging in front of the camera. He conveys an earnest passion and accessible demeanor that broadcasters and stem cell advocates alike exploited to full advantage. With his camera-readiness, Keirstead quickly found himself the poster scientist for everything stem cells could do. "Everybody wanted me to come and show my running rat video," he says, recounting how he found himself catapulted smack into the middle of the political and ethical debates over stem cell research.

Testifying before Congress, giving keynote speeches at investor conferences, and addressing hundreds of bioethicists are not activities most academic scientists expect once they have earned their Ph.D.'s. But Keirstead, gregarious by nature, welcomed the role and relished both the attention and the responsibility that it entailed. Like a child given a front row seat to a favorite show, he can hardly contain his giddiness even a decade later at finding himself at the vanguard of what he considers one of the most important revolutions in modern medicine.

That's because in January 2009, the treatment that Keirstead had squirted into the spinal cord of his paralyzed rats became the basis of the first embryonic stem cell–based therapy approved for testing in a human clinical trial.

★ ★ ★

If the stakes for the pharmaceutical industry are high, they are astronomically greater for the Food and Drug Administration and for Geron in conducting the first tests of a human embryonic stem cell therapy. With no precedent of any kind as guidance, the FDA and Geron started from scratch to set bars on safety and efficacy that were, to the best of their knowledge, sufficient. But without any prior experience, there was really no way they could know.

Hanging in Keirstead's office is a framed letter that he can now laugh over. It's from a reviewer of a scientific journal who gave Keirstead a brutally frank assessment of the paper the young researcher had submitted describing his success in creating a pure population of nerve-restoring cells from embryonic stem cells. "If these findings are true, this would be the most meaningful discovery in the stem-cell field," scoffed the reviewer, an expert in neurology. "But I simply don't believe it."

Neither did much of the rest of the spinal cord research community. By the 1990s, when Keirstead joined the faculty at the University of California, Irvine, and began his studies on repairing spinal cord injury, researchers had accepted that much of the damage arising from trauma to the cord is caused by the body's own response to the insult. In most cases, the injury itself does not sever the important rope of nerves running alongside it. The primary destruction is caused by the body's own alarmed reaction to the trauma. Immune cells, toxic agents, and inflammatory compounds released at the injury site end up destroying four to five times more nerve tissue than the initial lesion itself, breaking up delicate networks of neural connections. "Spinal cord injury is so complex; it's such a mess in there," Keirstead says of the frenzied tangle of tissues, cells, and chemicals that flare up after a trauma. Most experts in the field were focused on finding ways to reconnect these lost nerve fibers, and building bridges to restore the long, fingerlike projections of nerve cells, called axons, that extended from the spinal cord to the outermost extremities."

Keirstead, however, had another idea. After a four-year stint at the University of Cambridge studying a nerve cell known as the oligodendrocyte, he was convinced that there was another way to combat the chaotic destruction occurring in the minutes and hours following a spinal cord injury. In order to communicate, nerve cells rely on an insulating factor known as myelin. In exactly the same way that the electrical current in a wire is shielded within a thin tube of rubber, nerve cells are encased in myelin, and the coating improves their ability to conduct the electrical impulses that instruct a nerve to, for example, contract or stretch a muscle. In the frantic aftermath of a blow to the spinal cord, the body's immune cells begin to melt away this covering, stripping nerve cells of their ability to conduct messages and function normally. Without myelin, the signals instructing muscles to move stop dead in their tracks. The result is paralysis. And the cells that produce this myelin are none other than the oligodendrocytes, the tissue with which Keirstead had become quite experienced. He believed that if only there were a way to provide myelin-producing cells to a damaged spinal cord exactly where and when it needed them most, some of the function that was normally lost might be saved.

"Everyone in the field was trying to get axons of the spinal cord to regrow," he explains. "What I wanted to do was instead of regrowing the spinal cord, to patch it up." It was 2000, two years after Jamie Thomson had isolated the first human embryonic stem cell lines, and Keirstead knew how he would make his oligodendrocytes: He would generate them from embryonic stem cells.

Getting his hands on the cells would prove to be a feat almost as challenging as the experiment he was proposing. At the time, only a handful of investigators besides Thomson had generated human embryonic stem cells—Dickey-Wicker ensured that only those with deep private pockets were able to invest in the emerging technology. And because he wanted to eventually test his idea in people, Keirstead knew he would need the highest-quality cells he could find.

So he knocked on the door of the biotech company Geron, which at the time was the primary licensee of the highly coveted Thomson lines. Or rather, he slipped in the back door by calling in a favor from an old colleague from his Cambridge days, asking her to get him an introduction to the company as a guest speaker or a panelist, or anything that would give him the opportunity to present his idea to Geron's executives. It worked, and once Keirstead was invited to make his case, he wasted no time in making his intentions known. "I said, 'Give me some cells and money to make the world's first high-purity product in a commercially viable and clinically viable manner.'"

Geron bought the pitch, taking a gamble that the brash young scientist, who had just been given his own laboratory at Irvine, was onto something.

From Geron's perspective, Keirstead's treatment was too good to pass up. Having funded both Thomson and Gearhart in their efforts to generate the first human embryonic stem cell lines, the company was eager to affirm its position as the industrial leader in the stem cell space by being the first to take a stem cell–based therapy to patients. And having fought, since that very first congressional hearing in 1998, for the need for stem cell research to begin with, it seemed only right for Geron to be the first to approach the FDA with proof of the potential it had promised.

Biologically, the spinal cord therapy made sense as well. "Oh, sure, there were lots of considerations that went into our decision for a spinal cord injury trial," says Jane Lebkowski, the company's chief scientific officer. "When we first began thinking about trying to develop an embryonic stem cell therapy, we wanted an indication of where we would have to inject a relatively small number of cells. We didn't want to, in our very first clinical trial with these cells, have to manufacture a billion of them in order to do the study."

Keirstead's therapy fit that bill. His treatment consisted of using ES cells and transforming them into oligodendrocyte precursors. These would then be injected into the site of the spinal cord injury, where they would quickly adapt to their new surroundings and start blending in with the other cells to churn out the myelin that the nerves cells needed so badly. "We weren't asking that all of the circuitry in the spinal cord be reestablished; we were looking at these cells to do two things—produce nerve growth factors and remyelinate neurons," explains Lebkowski. "These are in some ways nursing activities. What we were looking at doing was protecting what was there already, so we thought the mechanism of action of these cells was probably a little easier to fulfill."

And it certainly looked promising, not only in Keirstead's walking rats but in additional animal studies that Geron conducted on its own to confirm and refine his results.

Both Keirstead and Geron, however, turned out to be a bit naive when it came to understanding what it would take to actually test the idea in a human patient. Having taken every twist and turn and complied with every request of the FDA to ensure that the therapy was safe and had reasonable expectation of helping patients, Keirstead still shakes his head over how involved the process turned out to be. "Boy, it's a longer road, and more difficult than people realize," he says of taking the first embryonic stem cell–based treatment from the lab bench to the clinic.

Over the course of six years, the company held numerous meetings with the FDA to both establish and meet standards for ensuring the safety of the stem cell–derived therapy that no one had ever had to fulfill before. The first challenge involved Keirstead's radical thesis itself—no one had ever shown that remyelinating spinal cord neurons could reduce the impact of the injury. Sure, it worked in rats, but would the same process apply to the much larger and potentially more complex human spinal cord? The agency called in neurologists to their Bethesda campus, consulted with other government experts at the National Institutes of Health, and learned that the answer was . . . maybe.

Then there were the embryonic stem cell issues. "This was the world's first high-purity stem cell derivate of any type," says Keirstead. One of the signatures of embryonic stem cells is the fact that they form tumors, which is actually a criterion that defines a pluripotent stem cell as a stem cell—these cells must have the ability to form a teratoma, a tumor containing all three germ layers (endoderm, mesoderm, and ectoderm). In a cell therapy, however, teratomas are the last thing you want. And while the Geron concoction of oligodendrocyte cells were not stem cells, per se, there was always the chance that an undifferentiated stem cell or two could have snuck into the batch. And all it takes is one such pluripotent cell to seed a teratoma. "No scientist or medical doctor in their right minds would ever think of putting an undifferentiated, tumor-forming stem cell into a human," says Keirstead. He argues that the risk of the oligodendrocyte cell preparation containing any such undifferentiated stem cells is extremely small, and that in reality, the chance of a tumor forming from this mixture is only as likely as the rate of tumors that normally develop from oligodendrocyte cells in the body, which is vanishingly small. "Because these cells come from a population of cells that can make tumors does not mean that the differentiated cell type maintains that capability," he says.

But that wasn't enough for the FDA. Without any prior database of studies and experiments with cells made from ES cells, agency officials were understandably nervous about even that small chance that the stem cells' tumor-forming tendencies would emerge again.

For Geron, that meant deliberately spiking their oligodendrocyte solution with undifferentiated cells and observing whether these intruders blossomed into teratomas.

And the caution proved well founded.

Eight months after the trial was approved, new data showed that animals injected with the cells developed a higher than expected number of odd rosettes at the injury site. These cysts of cells did not divide, and they did not migrate out of the spinal cord area to other parts of the body, says Lebkowski. Nor were they teratomas, suggesting that they were not, as the agency and company feared, undifferentiated stem cells that had snuck into the preparation. Still, the cysts were worrying and outside of the norm, so the agency suspended the trial. The FDA asked Geron to conduct additional studies to determine exactly what the cysts were and whether they would affect the function of the transplanted cells.

Nearly two years later, after countless additional tests that finally satisfied the agency, the FDA removed the hold and allowed Geron to proceed with its original plans to test the first stem cell–sired nerve cells in patients.

★ ★ ★

Almost universally, stem cell experts are dubious about how useful the Geron trial will be. Someone, they acknowledge, had to go first, and in doing so, Geron is laying critical groundwork for the next groups who will be ready to test their embryonic stem cell– or even iPS cell–based treatments in human patients. But as the first, Geron is in a catch-22 with progress—with advances in the field piling up nearly as quickly as they can be published in scientific journals, the company's cell preparation methods, already six years old, are outdated. The treatment that the

FDA approved, for example, is based on one of James Thomson's original human embryonic stem cell lines—at the time, the only cells that Geron and Keirstead had access to. That means the cells were grown on a feeder layer of mouse skin cells, just as Martin Evans's and Gail Martin's were in 1981. The mouse cells provided the nutrients and nurturing factors that helped stem cells to thrive, but can also introduce potential animal pathogens into the human cell solution. While the mouse cells are prepared under strictly regulated manufacturing requirements for sterility and safety, most researchers, including Keirstead, now grow only human stem cells using feeder-free methods. Any change to Geron's original protocol, however, would require the company to submit a new application with the new method.

Still, with every trial that the FDA reviews, its comfort with the standards and precedents it sets grows. That's something that Advanced Cell Technology (ACT), of Marlborough, Massachusetts, is exploiting with its own stem cell trial. ACT's is only the second such therapy that the agency has approved, and chief scientific officer Robert Lanza says ACT has already benefited from the progress made since Geron submitted its application—using more advanced imaging techniques, for instance, ACT scientists can detect a single undifferentiated cell in a teeming population of one million cells.

And to ensure that the stem cell therapy is as safe as possible, ACT chose to develop a treatment for retinal diseases such as macular degeneration and retinitis pigmentosa, in which the light-sensing photoreceptor cells in the retina begin to deteriorate. The retina is an ideal location for a novel therapy, because its cavity is an enclosed space, where cells rarely migrate in or out. It is also walled off from the immune system, so transplanted cells that are not immunologically matched to the host are safe from the body's cellular defenses. ACT's therapy is an injection of retinal pigment epithelium cells that are grown up from human embryonic stem cells. In treated animals, says Lanza, the thinning retinal cells in a diseased eye are built back up to a normal thickness of five to seven layers of cells.

"Whoever goes in there first has an incredible responsibility," he says of the pressure associated with the first stem cell–based human trials. "I can't tell you how many nights I've lost sleep thinking, 'Is there anything we're missing?'"

Regardless of the outcome of either the Geron or the ACT trials, the human studies will have been both necessary and worthwhile. "We have learned an awful lot, and asked a lot of questions," says Lebkowski of the process so far. "There are not studies that we can do in animal models that will say these cells are a hundred percent safe in humans. We can't do that until we move into the clinic. We can never be sure until we test this out in humans."

And no matter the result, those in the field are hoping that the research will only continue to grow and evolve. "This field has moved faster than we could have hoped," says Melton. "Because this is a very rich biology. It is going to change biomedicine."

Epilogue

Around noon on March 9, 2009, President Barack Obama signed a much anticipated executive order that lifted an eight-year stranglehold on stem cell research in the United States. Among those in attendance at the White House to witness the moment were familiar faces in the stem cell debate—patient advocates, some of the field's most stalwart scientific champions, as well as its rising stars. Harold Varmus. James Thomson. John Gearhart. Shinya Yamanaka. Irving Weissman. Susan Solomon. Zach Hall.

In explaining his reasons for signing the order, President Obama recognized the larger concern that had loomed over stem cell policy for nearly a decade—that politics and ideology had hampered the progress of science. "In recent years, when it comes to stem cell research, rather than furthering discovery, our government has forced what I believe is a false choice between sound science and moral values," he said. "In this case, I believe the two are not inconsistent. I believe we have been given the capacity and will to pursue this research—and the humanity and conscience to do so responsibly."

Doug Melton was not in Washington, choosing instead to take in the moment in Boston, in the lab with his students. "I bought a cake and some chocolate eggs, and we had a big celebration with thirty people," he tells me when I call him a few hours after the address. "I feel absolutely delighted and relieved."

One of the first things Melton mentions is the freedom he now feels upon being released from his regulatory cage. "Looking back, I realize how restrained and constrained we were by working in a silo imposed on us by the previous administration," he says. "Science is best done when there is an open exchange of ideas, reagents, and data. We were segregated from that."

His in-box, he said, was already full with even more requests for a few vials of his seventy human embryonic stem cell lines, which would eventually be added to the NIH registry. Many of the writers had been waiting for eight years to send those requests. There were a lot of questions to answer, and there was a lot of science to do.

★ ★ ★

Sam Melton, Jordan Klein, and Katie Zucker are in their early twenties now—young adults, with a strong sense of what President Obama's action means for them. For them, the address was the second such speech from a chief executive on stem cells that they had witnessed. They have lived through the political push-and-pull over the research, and the ups and downs of the regulatory seesaw. What happens to stem cell research will have a greater affect on them than on any scientist or legislator. They are, in essence, relying on and growing up alongside the science, maturing and developing just as it is.

In one way or another, they will be the first to benefit from the work researchers begin today. It might be in the form of a drug that comes from studying how beta cells made from embryonic stem cells lose their ability to make insulin. They might be spared having to become volunteers in testing a potentially toxic insulin-regulating compound, because it has already been screened with their iPS-derived cells and deemed unsafe for further development. Or maybe, just maybe, they could be the first recipients of an injection of beta cells grown up in the lab from stem cells.

They are realistic enough not to expect too much too soon. But they have also been following the research closely enough to know that something has to change.

Looking like a younger version of his father, Sam, unlike others his age, is accustomed to planning ahead—he has to be, in order to ensure that he adjusts his insulin levels according to how his body is responding. "If I were sick or didn't want to eat, or a friend asked if we wanted to play basketball, it caused chaos in terms of adjusting for the extra activity," he tells me. He is also accustomed to filling his pockets with Snickers and juice boxes, just in case his blood sugar levels drop.

He knows what his father has sacrificed in order to find a cure. He knows how hard his mother has worked to ensure that his life is as normal and unaffected by his disease as it can possibly be, despite the blood checks and the insulin shots. He knows the insults critics have leveled at his father, and what they think of the work he is doing with stem cells. He knows this because he was there, in the back of the room, when those opposed to embryonic research called his father an abomination. "It definitely made me angry. To me, there is nothing better than his research."

It's the same for Katie, recently graduated from New York University, having lived with diabetes for so long that she cannot imagine life without it and fighting just as long for the research that could lead to a cure. She was in the room, often curled up on the couch or sitting at the dinner table, when her parents gathered California's leading stem cell scientists at their home to figure out a way around the restrictive federal policy. In eighth grade, when her class was instructed to draw a political cartoon, she sketched a giant trash bin with a little stem cell sitting on its edge, with a caption that read, "I could have cured cancer."

As a twelve-year-old, Katie traveled to Washington to describe to members of Congress what it was like to live with diabetes. She couldn't appreciate it at the time, but she was there to fight a bill that would criminalize research on embryonic stem cells, the very cells that just might let her live without carrying around a bag full of needles and insulin and a glucose monitor. Nervous, she was afraid to say anything until one representative began telling her parents how the lives of the embryos had to be preserved. Confused, she finally spoke up. "What about my life?" she asked. "Doesn't that count for something?"

Jordan Klein was there too, to fight the same bill. He told senator John Kerry about what would happen if he slipped up and lost track of his blood sugar levels. He described the kidney failure, the blindness, the amputations, and the pain that would tear through his body if he forgot to do his blood checks or failed to give himself insulin when he needed it. After hearing his son relate all of this, his father had had second thoughts about bringing Jordan to Capitol Hill. Maybe, he told his son, it was too much. Maybe he shouldn't be doing this. "Don't worry," Jordan told his dad. "Life is life. Everyone is dying; I'm just dying a little faster."

Tall and lanky, with a head of tousled brown hair, Jordan knows what it's like to wake up every morning feeling as if his tissues have turned to stone, his body flooded with ketones that built up during the night. Growth hormone, secreted more copiously while he sleeps, sends his blood sugar levels through the roof, making mornings a challenge. For a while, he was afraid to go to sleep at night, fearing he would not wake up the next day.

"When I was younger, I definitely hoped for a cure soon," he admits. "But now . . . I'll just be ecstatic when it happens; I'd be so happy. But I'm not crossing my fingers or anything."

The caution seems odd for someone so young, someone who has seen a lot of hope in his short lifetime, from breakthroughs in the lab to less dramatic but still undeniable progress in policy. But it's a deliberation that comes from experience, from knowing victory often comes with a price.

Following the relief from President Obama's expansion of federally eligible stem cell studies, Jordan, Katie, and Sam were reminded yet again of how fleeting such triumphs can be in a science that remains controversial because of its embryonic origins. In a 2010 lawsuit reminiscent of the original one filed a decade earlier, Nightlight Christian Adoptions, along with some researchers focused on stem cells, sued the Obama administration for similar transgressions, alleging that any government grants for embryonic stem cell studies violated the Dickey-Wicker amendment.

And in a shock to the stem cell community, a federal judge agreed with them and forced the NIH to temporarily suspend its grants for embryonic stem cell studies. Even more astounding, the judge's decision appeared to preclude government support of all embryonic stem cell studies, including those that President Bush had allowed in his 2001 executive order. After an appeal by the White House, an appellate court issued a stay on the ban, but the suit will likely take another year to resolve in the courts, keeping embryo research on unsteady legal footing.

It's a reminder to the next generation of both scientists and patients that the fight over stem cells is hardly over, even with the existence of iPS cells. Before these newer versions can prove themselves equal to or even better than their embryonic predecessors, they will have to be compared side by side with them. And that means that at least for the time being, embryonic cell lines will continue to be a necessary part of stem cell studies, with all the political uncertainty and legal tensions that entails. The only resolution to the back-and-forth would be a law that establishes, with the same authority as Dickey-Wicker, what Harold Varmus had asked the Health and Human Services attorney to determine—that while Dickey-Wicker prohibited government funding of the creation of human embryonic stem cell lines, it did not preclude the NIH from offering grants to those who wanted to study the lines once they existed. Such legislation has been proposed in the House, several times, but like any issue before Congress, passing the bill is as much a matter of political agendas and motives as it is an issue of merit or urgency.

In the meantime, like any patient living with a chronic disease, Sam, too, is thinking about the cure. "I remember asking my parents when there would be a cure," he says, admitting it was an abstract concept for him then that has only recently become more tantalizingly concrete. For now, as he appreciates what it would take to find a permanent solution for diabetes, he, like his father, is convinced that stem cells will play some role in changing the way the disease impacts his life, even if it doesn't relieve him of his daily insulin shots. "I definitely think stem cells will contribute," he says.

He, along with the rest of world, is waiting to see how.

GLOSSARY

blastocyst. The four- to five-day stage of a developing embryo, containing the inner cell mass and the trophoblast, which goes on to become the placenta.

differentiate. To develop and specialize into a specific type of cell.

ectoderm. One of three primitive germ layers of embryo cells. It generates skin, hair, nails, epithelium, sense organs, mouth, and nervous tissue.

embryonal carcinoma (EC) cell. An abnormally growing cancer cell from the embryo that divides indefinitely.

embryonic stem (ES) cell. A stem cell obtained from an embryo, capable of generating the three basic germ layers (endoderm, ectoderm, and mesoderm), which create the more than two hundred different cell types in the human body. See also **stem cell**.

endoderm. One of three primitive germ layers of embryo cells. It generates the lining of the eustachian tube, trachea, lungs, gut, bladder, and vagina.

induced pluripotent stem cell (iPS). A stem cell made by reprogramming an adult cell (usually from the skin) back to its embryonic state using four transcription factors (genes), or by using their RNA equivalents. See also **stem cell**.

inner cell mass (ICM). The group of cells that form in the interior of a blastocyst and give rise to the embryonic stem cells.

mesoderm. One of three primitive germ layers of embryo cells. It generates muscle, cartilage, bone, bone marrow, blood vessels, kidneys, sex organs, and other tissues.

multipotent. Capable of developing into a limited number of the body's different cell types, such as bone marrow, which generates the blood and immune cells.

nuclear transfer, or **somatic cell nuclear transfer.** The process through which an adult cell is cloned by inserting its nucleus into an egg that has had its own nucleus removed. This is the procedure that resulted in Dolly, the first cloned mammal.

pluripotent. Capable of developing into nearly all of the body's different cell types. Embryonic stem cells are pluripotent, since they can create all but the placental cells of an embryo.

stem cell. A cell with the ability to develop into a variety of cell types.

therapeutic cloning. Nuclear transfer with the sole purpose of generating patient-specific stem cells by cloning that patient's cells.

trophectoderm. The exterior layer of cells of a blastocyst. It gives rise to the placenta.

unipotent. Capable of developing into only a single type of cell.

SOURCES

Preface

Interviews

Sinisa Hrvatin, November 19, 2009, Harvard University, Cambridge, Mass.; Robert Lanza, June 11, 2010, Marlborough, Mass.

Like any other cells Robert Lanza, J. Gearhart, B. Hogan, D. Melton, R. Pedersen, E. D. Thomas, J. Thomson, and M. West, eds., *Essentials of Stem Cell Biology* (Burlington, San Diego, London: Academic Press, 2006), xix.

So too do the recently isolated induced pluripotent stem (iPS) cells Kazutoshi Takahashi and Shinya Yamanaka, "Induction of Pluripotent Stem Cells from Mouse Embryonic and Adult Fibroblast Cultures by Defined Factors," *Cell* 126 (2006): 663–676; Kazutoshi Takahashi, K. Tanabe, M. Ohnuki, M. Narita, T. Ichisaka, K. Tomoda, S. Yamanaka, "Induction of Pluripotent Stem Cells from Adult Human Fibroblasts by Defined Factors," *Cell* 131 (2007): 861–872.

It's been only a decade since James Thomson, J. Itskovitz-Eldor, S. S. Shapiro, M. A. Waknitz, J. J. Swiergiel, V. S. Marshall, J. M. Jones, "Embryonic Stem Cell Lines Derived from Human Blastocysts," *Science* 282 (1998): 1145–1147.

the defective nerves that fail to stimulate muscles in patients with amyotrophic lateral sclerosis (ALS) John T. Dimos, K. T. Rodolfa, K. Niakan, L. M. Weisenthal, H. Mitsumoto, W. Chung, G. F. Croft, et al., "Induced Pluripotent Stem Cells Generated from Patients with ALS Can Be Differentiated into Motor Neurons, *Science* 321 (2008): 1218–1221.

toxic compound released by the glial cell Francesco Paolo Di Giorgio, G. L. Boulting, S. Bobrowicz, K. C. Eggan, "Human Embryonic Stem Cell-Derived Motor Neurons Are Sensitive to the Toxic Effect of Glial Cells Carrying an ALX-Causing Mutation," *Cell Stem Cell* 3 (2008): 637–648.

as a matter of debate in Congress Cloning and stem cell research have been the subject of special hearings in both the House and Senate, beginning December 2, 1998, Subcommittee of the Committee on Appropriations, United States Senate, Special Hearing Stem Cell Research, http://www.access.gpo.gov/congress/senate.

even the courts *Sherley et al. v. Sebelius et al.* (U.S. District Court for the District of Columbia, Case 1:09-cv-01575-RCL, August 23, 2010).

legislated a ban Public Law 104-99, 104th Congress (January 26, 1996).

President George W. Bush also acted, issuing an executive order President George W. Bush statement from Crawford, Texas (August 9, 2001), http://www.edition.cnn.com/2001/allpolitics/08/09/bush,transcript/index.html; Bush Executive Order no. 13,435, *Expanding Approved Stem Cell Lines in Ethically Responsible Ways* (June 22, 2007), http://www.archives.gov/federal-register/executive-orders/2007.html.

after President Obama removed these restrictions President Barack Obama Executive Order no. 13,505, *Removing Barriers to Responsible Scientific Research Involving Human Stem Cells* (March 9, 2009), http://www.archives.gov/federal-register/executive-orders/2009-obama.html.

Chapter 1

Interviews

Jordan Klein, June 19, 2010, Portola Valley, Calif.; Robert Klein, January 11, 2010, Portola Valley, Calif.; Douglas Melton, January 19, 2009, and November 19, 2009, Harvard University, Cambridge, Mass.; Gail O'Keefe, January 21, 2009.

Chapter 2

Interviews

John Gurdon, April 28, 2010; Rudolf Jaenisch, March 19–23, 2006, Knight Journalism Fellowship Stem Cell Boot Camp, Cambridge, Mass.; Gail Martin, September 8, 2009; Ian Wilmut, June 20, 2006, and May 6, 2010, New York.

Wilmut first announced Dolly's existence Ian Wilmut, A. E. Schnieke, J. McWhir, A. J. Kind, K. H. S. Campbell, "Viable Offspring Derived from Fetal and Adult Mammalian Cells," *Nature* 385 (1997): 810–813.

Born in Hampton Lucy Ian Wilmut and Roger Highfield, *After Dolly: The Uses and Misuses of Human Cloning* (New York: W. W. Norton & Company, 2006), 43.

Wilmut acknowledges a fondness Wilmut and Highfield, *After Dolly*, 15.

in 1982, his director abruptly ended Wilmut and Highfield, *After Dolly*, 50.

endowing the animals with the ability to produce a pharmacopoeia of milk Stephen Day, "Milking Them for All They're Worth," *Independent*, September 2, 1996.

and a 1958 paper by Sir John Gurdon M. Fischberg, J. B. Gurdon, T. R. Elsdale, "Nuclear Transplantation in *Xenopus laevis*," *Nature* 182 (1958): 64–65.

the pair reported Robert Briggs and Thomas King, *Journal of Morphology* 100 (1957): 269–312.

extended the work to the most mature cells transplanted at the time John Gurdon et al., *Nature* 210 (1966): 1240–1241.

Walter Cronkite made the trip Sir John Gurdon, interview by Harry Kreisler, March 16, 2006, *Conversations with History*, Institute of International Studies, University of California, Berkeley.

Gurdon's own students John Gurdon, *Nature Medicine* 15 (2009): 1141–1144.

scientist in Texas who had more or less successfully cloned a cow Wilmut and Highfield, *After Dolly*, 89.

presented with what he later called a "crazy" assignment Ricki Lewis, "A Stem Cell Legacy: Leroy Stevens," *Scientist* 14 (2000): 19–22.

managed to grow up stem cells Martin Evans et al., *Nature* 292 (1981): 154–156; Gail Martin, *Proceedings of the National Academy of Sciences* 78 (1981): 7634–7638.

found themselves picking up where Gurdon left off Wilmut and Highfield, *After Dolly*, 93.

The result were Megan and Morag Wilmut and Highfield, *After Dolly*, 97.

In the dead of winter Wilmut and Highfield, *After Dolly*, Chapter 3.

Wilmut says it was he Wilmut and Highfield, *After Dolly*, 26–27.

An autopsy found Jim Giles and Jonathan Knight, *Nature* 421 (2003): 776.

a stunning array of other mammals Jose B. Cibelli et al., *Nature Biotechnology* 20 (2002): 13–14.

Chapter 3

Interviews

R. Alta Charo, September 30, 2009, and May 14, 2010; Jay Dickey, September 10, 2010; Brigid Hogan, February 8, 2010; Dr. Bernard Lo, January 15, 2010, San Francisco, Calif.; Thomas Murray, December 15, 2009; Carol Tauer, October 15, 2009; Harold Varmus, November 25, 2009, New York.

President Clinton quickly diverted The White House Office of the Press Secretary, "President Clinton Announces Cloning Prohibition Act of 1997," http://clinton3.nara.gov/New/Remarks/Mon/19970609-16602.html.

addressing Congress Subcommittee of the Committee on Appropriations, United States Senate, Special Hearing Stem Cell Research, December 2, 1998, http://www.access.gpo.gov/congress/senate.

President Clinton lifted Robin Toner, "Clinton Orders Reversal of Abortion Restrictions Left by Reagan and Bush," *New York Times*, January 23, 1993.

Clinton wrote his decision into law National Institutes of Health Revitalization Act, Public Law 103-43, 103rd Congress (June 10, 1993), http://www.hhs.gov/ohrp/humansubjects/guidance/publiclaw103-43.htm.

It was a typically chilly February day Ronald Green, *The Human Embryo Research Debates: Bioethics in the Vortex of Controversy* (New York: Oxford University Press, 2001), Chapter 1.

Congress quickly convened D. C. Wertz, "Embryo and Stem Cell Research in the United States: History and Politics," *Gene Therapy* 9 (2002): 674–678.

EAB came back with its report http://bioethics.georgetown.edu/pcbe/reports/past_commissions/HEW_IVF_report.pdf.

Califano fell victim Edward Walsh, "Califano, Blumenthal Are Fired from Cabinet," *Washington Post*, July 20, 1979.

That was what a panel debating similar thorny issues *Report of the Committee of the Inquiry into Human Fertilisation and Embryology*, June 24, 2008, http://www.hfea.gov.uk/docs/Warnock_Report_of_the_Committee_of_Inquiry_into_Human_Fertilisation_and_Embryology_1984.pdf.

Human Fertilisation and Embryology Act in 1990 Human Fertilisation and Embryology Act 1990, http://www.legislation.gov.uk/ukpga/1990/37/contents.

In the end, the nineteen members voted *Report of the Human Embryo Research Panel*, National Institutes of Health, September 27, 1994.

to write legislation that followed Clinton's lead *The Balanced Budget Downpayment Act*, Public Law 104-99, Sec. 128, 104th Congress (January 26, 1996).

Chapter 4

Interviews

R. Alta Charo, September 30, 2009; Lynn Boehnlein, March 26, 2010, Madison, Wis.; Terry Devitt, March 26, 2010, Madison; Erik Forsberg, March 26, 2010, Madison, Wis.; Carl Gulbrandsen, March 25, 2010, Madison; Joseph Itskovitz-Eldor, March 16, 2010; Tenneille Ludwig, March 26, 2010, Madison, Wis.; Douglas Melton, January 28, 2010, Cambridge, Mass.; Roger Pedersen, April 26, 2010; James Thomson, March 25, 2010, Madison, Wis.

James Thomson is a ferociously private man Terry Devitt, "Reluctant Star," *On Wisconsin*, Spring 2008.

When researchers first identified National Human Genome Research Institute's summary of knockout mouse milestones, http://www.genome.gov/12514551.

researchers were beginning to appreciate Yuehua Jiang et al., *Nature* 418 (2002): 41–49. In the paper, lead author Catherine Verfaillie, of University of Minnesota, describes turning a group of adult stem cells into all three germ cell layers, implying that adult stem cells can be as versatile as embryonic stem cells. However, the results have been questioned because researchers have only been able to replicate parts of the study and have yet to repeat the key test of the pluripotentiality of adult stem cells—injecting the adult stem cells into a mouse embryo to verify that they can generate all of the body's major cell types. See Erika Check, "The Hard Copy," *Nature* 446 (2007): 485–486.

adult progenitors were proving Lanza, *Essentials of Stem Cell Biology*, Chapter 3.

a string of scientists had tried Davor Solter, *Nature Reviews Genetics* 7 (2006): 319–327.

In the 1980s, Andrews Miodrag Stojkovic et al., *Stem Cells* 27 (2009): 4–6.

Called immunosurgery Davor Solter and Barbara Knowles, *Proceedings of the National Academy of Sciences* 72 (1975): 5099–5102.

In the five-page publication James Thomson et al., *Proceedings of the National Academy of Sciences* 92 (1995): 7844–7848.

but even his article Associated Press, "New Gene Therapy Moves Ahead; Researchers Isolate Monkey's 'Parent' Cells," November 14, 1994.

The university is regularly among http://www.rsp.wisc.edu/fednih.html.

Years later, Thomson remarked Gina Kolata, "Man Who Helped Start Stem Cell War May End It," *New York Times*, November 22, 2007.

His three-page report Thomson et al. (1998).

Within a month Subcommittee of the Committee on Appropriations (1998), http://www.access.gpo.gov/congress/senate.

The nonprofit Wisconsin Alumni Research Foundation (WARF) Michael Penn, "Working Beyond Eureka!" *On Wisconsin*, Spring 2002, http://www .warf.org.

pulling in several billion each year Peter Gorner and Ronald Kotulak, "Play Doctor . . . or Playing God?" *Chicago Tribune*, December 9, 2001.

WARF would be seen Sheryl Gay Stolberg, "Patent on Human Stem Cell Puts U.S. Officials in Bind," *New York Times*, August 17, 2001.

Initially, it charged Marilynn Marchione, "UW Agency Sues Firm Over Stem Cells," *Milwaukee Journal Sentinel*, August 14, 2001.

the biotech company enjoyed exclusive rights Tom Abate, "Geron Target of Stem Cell Suit," *San Francisco Chronicle*, August 15, 2001.

thousands of genes had already been patented Scott LaFee, "Patent Offending," *San Diego Union-Tribune*, April 19, 2006.

foundation opted to patent WARF filed two patents related to Thomson's embryonic stem cell lines, generated from primate and human cells, U.S. Patent 5,843,780, http://patft.uspto.gov/netacgi/nph-Parser?Sect1=PTO1&Sect2= HITOFF&d=PALL&p=1&u=%2Fnetahtml%2FPTO%2Fsrchnum.htm&r= 1&f=G&l=50&s1=5843780.PN.&OS=PN/5843780&RS=PN/5843780

&RS=PN/5843780, and 6,200,806, http://patft.uspto.gov/netacgi/nph-Parser
?Sect1=PTO1&Sect2=HITOFF&d=PALL&p=1&u=%2Fnetahtml%2FPTO
%2Fsrchnum.htm&r=1&f=G&I=50&s1=6200806.PN.&OS=PN/6200806
&RS=PN/6200806.

Chapter 5

Interviews

Dr. James Battey Jr., June 18, 2010, Richard Doerflinger, June 23, 2010, Bethesda, Md.; Jay Lefkowitz, December 14, 2009, New York; Douglas Melton, November 19, 2009, Cambridge, Mass.; Harold Varmus, November 25, 2009, New York; Irving Weissman, January 15, 2010, Palo Alto, Calif.

He was against abortion "Under Fire, Bush Breaks Silence on Abortion," *Boston Globe*, January 21, 2000; "Bush Pressed for Abortion Stance," *Los Angeles Times*, January 21, 2000.

an act he viewed "Biomedical Research Goes Where Candidates Dare Not," *Washington Post*, October 29, 2000.

For three days Lefkowitz described the process behind Bush's stem cell decision in "Stem Cells and the President—An Inside Account," *Commentary*, January 2008.

becoming a critical political bellwether "Unexpected Priority: Stem Cell Research's Rise As a Test for Bush," *New York Times*, July 14, 2001.

giving Bush's decision a gravitas "The President's Decision: The Overview; Bush Gives His Backing for Limited Research on Existing Stem Cells," *New York Times*, August 10, 2001.

Bush was considering banning "Bush Aides Seek Compromise on Embryonic Cell Research," *New York Times*, July 4, 2001.

he has earned the nickname Viper "A Hard-Nosed Litigator Becomes Bush's Policy Point Man," *Washington Post*, April 30, 2002.

The suit was filed *Nightlife Christian Adoptions et al. v. Thompson et al.* (U.S. District Court for the District of Columbia, March 8, 2001).

set of guidelines evaluating incoming research proposals *Federal Register* 64, no. 231 (December 2, 1999), http://stemcells.nih.gov/staticresources/news/newsArchives/fr02de99-43.asp.

"Taxpayer funds should not underwrite" "Bush Caught in the Middle on Research on Stem Cells," *New York Times*, February 18, 2001.

all traipsed through the Oval Office "Bush 'Agonizing' Over Funding of Embryo Research," *Washington Post*, July 15, 2001.

Cynics saw the ultimate spin campaign Richard Cohen, "The Stem Cell Student," *Washington Post*, July 17, 2001.

Complicating matters Robert Pear, "Several G.O.P. Senators Back Money for Stem Cell Research," *New York Times*, June 19, 2001.

Within the administration itself Ceci Connolly and Rick Weiss, "Stem Cell Research Divides Administration," *Washington Post*, June 12, 2001.

"There is a law that has been passed by Congress" Dana Milbank, "In Bush Cabinet, It's Both Advise and Dissent," *Washington Post*, March 10, 2001.

NIH's planned meeting to review the first grant proposals Rick Weiss, "Bush Administration Order Halts Stem Cell Meeting," *Washington Post*, April 21, 2001.

ended up making an important decision Harold Varmus, *The Art and Politics of Science* (New York: W. W. Norton & Co., 2009), 219.

"No Federal funds were used" Harold Varmus, U.S. Senate Hearing Subcommittee of Committee on Appropriations, January 26, 1999.

June letter to the Culture of Life Foundation Robert Pear, "Bush Administration Is Split Over Stem Cell Research," *New York Times*, June 13, 2001.

"To date," the report concluded National Institutes of Health, *Stem Cells: Scientific Progress and Future Research Directions*, June 2001.

"Why shouldn't these embryos" Orrin Hatch, U.S. Senate Hearing Subcommittee of the Committee on Appropriations, July 18, 2001, http://frwebgate.access.gpo.gov/cgi-bin/getdoc.cgi?dbname=107_senate_ hearings&docid=f:77047.pdf.

"We do not kill" Richard Doerflinger, ibid.

"Stem cell research facilitates life," Ceci Connolly, "Conservative Pressure for Stem Cell Funds Builds," *Washington Post*, July 2, 2001.

Thirty-eight House Republicans ibid.

"The federal government cannot morally look" Ceci Connolly, "3 GOP Leaders Warn Bush on Stem Cell Studies," *Washington Post*, July 3, 2001.

scientists at the Jones Institute for Reproductive Medicine Sheryl Gay Stolberg, "Scientists Create Scores of Embryos to Harvest Cells," *New York Times*, July 11, 2001.

researchers at Advanced Cell Technology "Company Using Cloning to Yield Stem Cells," *New York Times*, July 13, 2001.

"Which one of my children" John Borden and Lucinda Borden, Hearing of U.S. House of Representatives Subcommittee on Criminal Justice, Drug Policy, and Human Resources of the Committee on Government Reform, July 17, 2001, http://frwebgate.access.gpo.gov/cgi-bin/getdoc .cgi?dbname=107_house_hearings&docid=f:77248.pdf.

a report by MIT's Rudolph Jaenisch *Science* 293 (2001): 1093–1098.

Nancy Reagan, having watched "Nancy Reagan Backs Research on Stem Cells," *Washington Post*, July 14, 2001.

"A free and virtuous society," "Remarks by John Paul," *New York Times*, July 24, 2001.

introduced legislation that would ban human cloning U.S. House of Representatives, Human Cloning Prohibition Act of 2001, H.R. 2505, July 31, 2001.

Chapter 6

Interviews

Nissim Benvenisty, March 4, 2010, New York; Thomas Cech, January 8, 2010; Chad Cowan, January 29, 2010; Carl Gulbrandsen, March 25, 2010, Madison, Wis.; John Gurdon, April 28, 2010; Jay Lefkowitz, December 14, 2009, New York; Douglas Melton, March 21, 2006, November 19, 2009, and June 10, 2010, Cambridge, Mass.; Roger Pedersen, June 30, 2009; Peter Van Etten, February 10, 2010; Laurie Zoloth, March 17, 2010.

Bush's speech http://usgovinfo.about.com/blwhrelease16.htm?p=1.

As more researchers voiced their skepticism Nicholas Wade, "Worried Scientists Are Told Ample Stem Cell Lines Exist," *New York Times*, August 25, 2001.

Just days before Bush left Brian Vastag, "Suddenly, 64 Stem Cell Lines," *Journal of the American Medical Association* 286 (2001): 1163–1164; Lana Skirboll, "The Development of a Stem Cell Policy in the US," presentation at European Commission meeting Stem Cells: Therapies for the Future? (December 18–19, 2001), http://ec.europa.eu/research/quality-of-life/stemcells/pro_skirboll_en.html.

"They're diverse, they're robust" Sheryl Gay Stolberg, "U.S. Concedes Some Cell Lines Are Not Ready," *New York Times*, September 6, 2001.

"I used to hold the view" Arthur Allen, "God and Science," *Washington Post*, October 15, 2000.

agreed to a collaboration Maya Pines, "Are Stem Cells the Answer?" *HHMI Bulletin*, March 2002.

Melton and the McMahons began "Stem-cell Science," *Harvard Magazine*, July–August 2004.

In March 2004, Melton's team published a paper Chad A. Cowan, I. Klimanskaya, J. McMahon, J. Atienza, J. Witmyer, J. P. Zucker, S. Wang, et al., *New England Journal of Medicine* 350 (2004): 1353–1356.

Chapter 7

Interviews

Paul Berg, January 15, 2010, Palo Alto, Calif.; Marie Csete, May 27, 2010; Larry Goldstein, August 6, 2009; Zach Hall, October 15, 2009, New York; Robert Klein, January 11, 2010, Portola Valley, Calif.; Ana Matosantos, April 8, 2010; Deborah Ortiz, November 5, 2009; Jeff Sheehy, August 8, 2009, and June 8, 2010; Alan Trounson, January 15, 2010, San Francisco, Calif.; Peter Van Etten, February 10, 2010; Irving Weissman, January 15, 2010, Palo Alto, Calif.; Jerry Zucker, February 4, 2010.

befriended a woman whose life Carl T. Hall, "One Man's Scientific Mission," *Los Angeles Times*, April 11, 2005.

he and BeVier went to the State of California Lisa Krieger and Paul Jacobs, "Financier Set to Oversee Stem-Cell Program," *San Jose Mercury News*, January 16, 2005.

Governor Gray Davis signed Gregg Jones, "Bill Boosting Stem-Cell Research to Be Signed," *Los Angeles Times*, September 22, 2002.

Together, Ortiz and the Zuckers Connie Bruck, "Hollywood Science," *The New Yorker*, October 18, 2004.

convening a group of genetic experts Paul Berg, "Meetings That Changed the World: Asilomar 1975: DNA Modification Secured," *Nature* 455 (2008): 290–291.

voluntary moratorium http://www.hms.harvard.edu/orsp/coms/government/The-Guidelines-Explained.pdf.

The funding would be split Carl T. Hall, "State Voters Strongly Backing Cell Research," *San Francisco Chronicle*, November 3, 2004.

influenced by Nancy's personal request "Nancy Reagan Backs Cloning for Research," *Washington Post*, February 7, 2003.

Nancy took her appeal public Laura Mecoy, "Nancy Reagan Presses the Fight," *Sacramento Bee*, May 8, 2004.

Californians were nearly split Keay Davidson, "Stem Cell Initiative Leads by Small Margin," *San Francisco Chronicle*, August 15, 2004.

People's Advocate and the National Tax Limitation Committee called for the disbanding Bob Egelko, "State Supreme Court Refuses to Hear Challenge to Stem Cell Research Program," *San Francisco Chronicle*, March 24, 2005.

intricate selection rules For text of Proposition 71, see http://www.cirm.ca.gov/AboutCIRM_Prop71.

the five-hour meeting For transcripts of CIRM meetings, see http://www.cirm.ca.gov.

proposed a bill that would change Bernadette Tansey, "Stem Cell Advocate Tries to Plug Prop. 71 Holes," *San Francisco Chronicle*, December 6, 2004.

survived on a start-up loan Carl T. Hall, "$5 Million Donation for Stem Cell Startup," *San Francisco Chronicle*, June 6, 2005.

bond anticipation notes Carl T. Hall, "Stem Cell Panel Goes to Work on Funding," *San Francisco Chronicle*, May 10, 2005.

award its first grants CIRM, "Stem Cell Institute Awards First Scientific Grants," April 10, 2006, http://www.cirm.ca.gov/PressRelease_041006.

awarded its first bona fide scientific grants CIRM, "$45 Million Headed for Stem Cell Research in California," February 16, 2007, http://www.cirm.ca.gov/PressRelease_021607.

"scientifically thin," Carl T. Hall, "Stem Cell Grants Come with Dash of Criticism," *San Francisco Chronicle*, March 31, 2007.

Chapter 8

Interviews

Hidenori Akutsu, October 22, 2009, Tokyo; John Dimos, August 6, 2009; Diana DeGette, August 28, 2009; Richard Doerflinger, June 23, 2010; Kevin Eggan, July 10, 2009, Barcelona, Spain; Susan Fisher, July 14, 2009, San Francisco, Calif.; John Gurdon, April 28, 2010; Dr. William Hurlbut, January 12, 2010, Palo Alto, Calif.; Insoo Hyun, April 15, 2010; Douglas Melton, January 28, 2010, Cambridge, Mass.; Renee Reijo Pera, January 12, 2010, Palo Alto, Calif.; Teruhito Wakayama, October 21, 2009, Kobe, Japan, and July 8, 1998, New York; Laurie Zoloth, March 17, 2010.

scientists, including Ian Wilmut Tim Friend, "Scientist Calls for Limited Human Cloning," *USA Today*, March 13, 1997.

Human Cloning Prohibition Act The White House Office of the Press Secretary, "President Clinton Announces Cloning Prohibition Act of 1997."

recommendations of the National Bioethics Advisory Commission (NBAC) National Bioethics Advisory Commission, "Human Stem Cell Research," September 1999, http://www.bioethics.gov/commissions/.

Congress, however, never passed Human Cloning Prohibition Act of 2001, H.R. 2505, 107th Congress; Human Cloning Prohibition Act of 2003, H.R. 534, 108th Congress; Human Cloning Prohibition Act of 2009, H.R. 1050, 111th Congress. Three times the House passed by a comfortable margin the Human Cloning Prohibition Act, only to have the legislation stall in the Senate.

offered to clone members Reuters, "Swiss Group Launches Firm to Market Human Cloning," *San Jose Mercury News*, June 9, 1997.

"The less everybody knows" Sharon Kirkey, "Richard Seed: The Face of Human Cloning," *Ottawa Citizen*, October 15, 1999.

Korean infertility researchers Sheryl Wu Dunn, "South Korean Scientists Say They Cloned a Human Cell," *New York Times*, December 17, 1998.

failed, however, to follow basic scientific protocol Rick Weiss, "Cloned Human Embryo Created, South Korean Researchers Say," *Washington Post*, December 17, 1998; Associated Press, "Panel Casts Doubt on Human Cloning Claim," *New York Times*, January 29, 1999.

Representative John Greenwood, a Republican, who called a hearing For transcript, see "Issues Raised by Human Cloning Research," Hearing before Subcommittee on Oversight and Investigations of the Committee on Energy and Commerce, 107th Congress, March 28, 2001.

The university's program was the first "Harvard Stem Cell Researchers Granted Approval," *Harvard Gazette*, June 8, 2006; Gareth Cook, "Harvard Launches Effort to Clone Human Stem Cells," *Boston Globe*, June 7, 2006.

Across the country, at UCSF Carl T. Hall, "UCSF Resumes Human Embryo Stem Cell Work," *San Francisco Chronicle*, May 6, 2006.

both advised against federal support National Bioethics Advisory Commission, "Human Stem Cell Research," September 1999; The President's Council on Bioethics, "Human Cloning and Human Dignity: An Ethical Inquiry," July 2002; both at http://www.bioethics.gov/commissions/.

led efforts in Congress Stem Cell Research Enhancement Act of 2005, H.R. 810, 109th Congress; Stem Cell Research Enhancement Act of 2007, S. 5, 110th Congress. Both bills broadened the reach of federal funding to include generating human embryonic stem cells from excess IVF embryos. Both passed in the House and Senate.

"Repugnance is the emotional expression" Leon Kass, "The Wisdom of Repugnance," *New Republic* 216 (1997): 17–26. Kass declined to comment on his role in the PCB, referring me instead to his publications.

invitation for researchers to think "creatively" "Alternative Sources of Pluripotent Stem Cells," press teleconference transcript, May 12, 2005, http://www.bioethics.gov/commissions/.

council outlined four other ways President's Council on Bioethics, "White Paper: Alternative Sources of Pluripotent Stem Cells," May 2005, http://www.bioethics.gov/commissions/.

Rudolph Jaenisch, at MIT, put the idea to the test Alexander Meissner and Rudolph Jaenisch, "Generation of Nuclear Transfer-Derived Pluripotent ES Cells from Cloned Cdx2-Deficient Blastocysts," *Nature* 439 (2006): 212–215.

Commenting on the proposal Douglas A. Melton, George Q. Daley, Charles G. Jennings, "Altered Nuclear Transfer in Stem Cell Research—A Flawed Proposal," *New England Journal of Medicine* 351 (2004): 2791–2792.

proved that therapeutic cloning could actually cure a disease William M. Rideout III, K. Hochedlinger, M. Kyba, G. Q. Daley, R. Jaenisch, "Correction of a Genetic Defect by Nuclear Transplantation and Combined Cell and Gene Therapy," *Cell* 109 (2002): 17–27.

fused a human adult cell with a human embryonic stem cell Chad A. Cowan, J. Atienza, D. A. Melton, K. Eggan, "Nuclear Reprogramming of Somatic Cells After Fusion with Human Embryonic Stem Cells," *Science* 309 (2005): 1369–1373.

Chapter 9

Interviews

Jiho Choi, January 28, 2010, Cambridge, Mass.; Woo Suk Hwang, May 9, 2005, Seoul, Korea, May 20, 2005, Incheon, Korea, and September 30, 2005; Insoo Hyun, August 13, 2009; Kyu Won Jung, October 29, 2009, Seoul; Sung Keun Kang, May 9, 2005, Seoul; Ok Jae Koo, May 9, 2005, Seoul;

Byeong Chun Lee, May 9, 2005, Seoul; Ki Young Park, May 10, 2005, Seoul; Dr. Jeong Bin Yim, May 10, 2010, Seoul.

Publishing his breakthrough Woo Suk Hwang et al., *Science* 303 (2004): 1669–1674.

he had used the cloning process Woo Suk Hwang et al., *Science* 308 (2005): 1777–1783.

began asking some of Hwang's coauthors David Cyranoski, *Nature* 429 (2004): 3.

Hwang continued to deny Dennis Normile, *Science* 304 (2004): 945.

"deeply betrayed" by his colleagues' research David Cyranoski *Nature* 429, 490 (2004).

In Jung's process, which he described Kyu Won Jung, *American Journal of Bioethics* 6 (2006): W1–4

In June, he announced Associated Press, "Cloning Pioneer Plans to Open Stem Cell Bank," June 1, 2005.

it received so many requests Associated Press, "Patients Flood South Korean Stem Cell Site," October 31, 2005.

cryptic press release University of Pittsburgh, "University of Pittsburgh Researcher Ends Collaboration with South Korean Stem Cell Program," November 12, 2005.

"I know nothing" Jin-seo Cho, "All Ethical Codes Met, Says Hwang," *Korea Times*, November 14, 2005.

an e-mail arrived Sei Chong and Dennis Normile, *Science* 311 (2006): 22–25.

retreated to a Buddhist temple Cho, "All Ethical Codes Met, Says Hwang."

"I did this work without consulting" Tae-gyu Kim, "Member of Hwang Team Involved in Ovum Scandal," *Korea Times*, November 8, 2005.

producers interviewed women who claimed that they had sold their eggs "MBC Team Charges Stem Cell Research 'a Sham,'" *Chosunilbo*, November 28, 2005; David Cyranoski, *Nature* 438 (2005): 405.

Hwang publicly admitted to lying Associated Press, "South Korean Cloning Pioneer Resigns," November 24, 2005.

"being too focused on scientific development," ibid.

violent threats against producers at MBC "MBC Apologizes for Bullying in Hwang Expose," *Chosunilbo*, December 4, 2005.

eleven of the twelve sponsors of *PD Notebook* terminated their funding "MBC Pays for Scandals with Advertising Slump," *Chosunilbo*, December 1, 2005; "Fresh Hwang Report Lands MBC in Hot Water Again," *Chosunilbo*, December 2, 2005.

That came on December 4, 2005 As events unfolded from November to December 2005, the editors of *Science* posted and continued to update a detailed timeline of events in the Hwang scandal. The chronology is part of the collection of papers, retractions, and editorials concerning the episode, http://www.sciencemag.org/sciext/hwang2005/.

"Our research team made patient-specific embryonic stem cells," "Cell-Making Process Can Be Repeated," *Korea Times*, December 16, 2005.

2004 cell lines could not be confirmed Seoul National University Investigation Committee report, January 10, 2006.

"I feel so crushed and humiliated" Sang-hun Choe, "South Korea Scientist Contrite for Stem Cell Fraud," *International Herald Tribune*, January 12, 2006.

Hwang was brought to trial Sei Chong, "Investigations Document Still More Problems for Stem Cell Researchers," *Science* 311 (2006): 754–755.

on the shoulders of one man Dong-hee Wohn, "Journal Says It May Cancel Hwang Paper Unilaterally," *JoongAng Daily*, December 26, 2005.

independent lab confirmed Seoul National University Investigation Committee, "Verification that Snuppy Is a Clone," *Nature* 440 (2006): E2–E3.

Chapter 10

Interviews

Valerie Estess, September 11, 2009, October 1, 2009, and October 9, 2009, New York; Robert Farese, July 14, 2009, San Francisco; Christopher Henderson, October 9, 2009, New York; Thomas Jessell, October 1, 2009, New York; Scott Noggle, December 11, 2009, New York; Susan Solomon, June 24, 2009, New York, and June 18, 2010, San Francisco; Hynek Wichterle, October 9, 2009, New York.

"Some nights thinking about work" Jenifer Estess, *Tales From the Bed: A Memoir* (New York: Washington Square Press, 2005), 145–146. See in general for description of Jenifer's diagnosis with ALS, as told to her sister Valerie.

some of them stopped shivering Booma Yandava et al., *Proceedings of the National Academy of Sciences* 96 (1999): 7029–7034.

march mouse embryonic stem cells through their various developmental stages Hynek Wichterle et al., *Cell* 110 (2002): 385–397.

Chapter 11

Interviews

Kevin Eggan, June 10, 2010, Cambridge, Mass.; Thomas Jessell, October 1, 2009, New York; Deepak Srivastava, December 14, 2009; Kazutoshi Takahashi, October 19, 2009, Kyoto, Japan; James Thomson, March 25, 2010, Madison, Wis.; Shinya Yamanaka, July 8, 2009, Barcelona, Spain, October 19, 2009, Kyoto, Japan, and June 16, 2010, San Francisco, Calif.

Recognized with a Lasker Award http://www.laskerfoundation.org.

which he dubbed NAT1 Shinya Yamanaka et al., *Genes and Development* 11 (1997): 321–333.

splice the gene into the genome of mouse embryonic skin cells Robert Davis et al., *Cell* 51 (1987): 987–1000; Harold Weintraub et al., *Proceedings of the National Academy of Sciences* 86 (1989): 5434–5438.

likened his scientific career to an *ekiden* Shinya Yamanaka, "Ekiden to iPS Cells," *Nature Medicine* 15 (2009): xi–xiv.

four factors absolutely required Takahashi et al., *Cell* 126 (2006): 663–676.

simultaneously announced their success in making human iPS cells Takahashi et al., *Cell* 131 (2007): 861–872; Junying Yu, M. A. Vodyanik, K. Smuga-Otto, J. Antosiewicz-Bourget, J. L. Frane, S. Tian, J. Nie, et al., "Induced Pluripotent Stem Cell Lines Derived From Human Somatic Cells," *Science* 318 (2007): 1917–1920.

conjuring up motor neurons Dimos et al., *Science* 321 (2008), 1218–1221.

Chapter 12

Interviews

Justin Annes, September 24, 2009, Cambridge, Mass.; Konrad Hochedlinger, January 28, 2010, Cambridge, Mass.; Hans Keirstead, April 20, 2010; Story Landis, Bethesda, Md.; Robert Lanza, June 11, 2010, Marlborough, Mass.; Jane Lebkowski, July 3, 2006, Toronto, and June 19, 2010, San Francisco; Ihor Lemischka, June 4, 2010; Ruth McKernan, January 27, 2010, Cambridge, Mass.; Douglas Melton, September 24, 2009, Cambridge, Mass.; Derrick Rossi, September 29, 2010; Lee Rubin, January 29, 2010, Cambridge, Mass.; Deepak Srivastava, October 4, 2010, San Francisco, Calif.

creating the first stem cell lines from two young boys with type 1 diabetes René Maehr, S. Chen, M. Snitow, T. Ludwig, L. Yagasaki, R. Goland, R. L. Leibel, et al., "Generation of Pluripotent Stem Cells from Patients with Type 1 Diabetes," *Proceedings of the National Academy of Sciences* 106 (2009): 15768–15773.

Yamanaka himself managed to swap out the viruses Keisuke Okita, M. Nakagawa, H. Hyenjong, T. Ichisaka, S. Yamanaka, "Generation of Mouse Induced Pluripotent Stem Cells Without Viral Vectors," *Science* 322 (2008): 949–953.

Melton replaced two of the four Danwei Huangfu, K. Osafune, R. Maehr, W. Guo, A. Eijkelenboom, S. Chen, W. Muhlestein, et al., "Induction of Pluripotent Stem Cells from Primary Human Fibroblasts with Only Oct4 and Sox2," *Nature Biotechnology* 26 (2008): 1269–1275.

the strategy not only worked Luigi Warren et al., *Cell Stem Cell* 7 (2010): 618–630.

beta cell has no adult stem cell Kristen Brennand et al., *PLoS Biology* 5 (2007): 1520–1529.

convinced it to produce insulin instead Qiao Zhou et al., *Nature* 455 (2008): 627–633.

turning human skin cells directly into heart muscle Masaki Ieda et al., *Cell* 142 (2010): 1–12.

Epilogue

Interviews

Jordan Klein, June 19, 2010, Portola Valley, Calif.; Douglas Melton, March 9, 2009; Sam Melton, January 21, 2009, Cambridge, Mass., and June 10, 2010; Katie Zucker, May 7, 2010, New York.

President Barack Obama signed a much anticipated executive order Executive Order no. 13,505, *Federal Register* 74, no. 46 (March 11, 2009). For a history of government stem cell policies, see http://stemcells.nih .gov/policy.

federal judge agreed with them and forced the NIH *Sherley et al. v. Sebelius et al.* (U.S. District Court for the District of Columbia, Case 1:09-cv-01575-RCL, August 23, 2010).

ADDITIONAL RESOURCES

California Institute for Regenerative Medicine

http://www.cirm.ca.gov/

One of the largest funders of research on stem cells, including embryonic, adult, and induced pluripotent stem cells. The site provides basic education about stem cells, as well as details about the latest studies in the field, funded by California citizens. Because the institute is public, state law requires that all of its meetings, with few exceptions, be open and documented for the public.

Harvard Stem Cell Institute

http://www.hsci.harvard.edu/

One of the first institutes to focus specifically on stem cell studies, HSCI continues to conduct cutting-edge research aimed at translating the basic science findings into clinical applications, from new drugs to the next generation of cell-based treatments for disease. The site is continually updated with research from the institute's diverse faculty.

National Institutes of Health

http://stemcells.nih.gov/

A good resource for the latest government policies on stem cell research, as well as basic information on stem cells and how they might be applied to treating disease.

New York Stem Cell Foundation

http://www.nyscf.org/

NYSCF was created in response to President Bush's restriction on federal funding of stem cell research in 2001, and has since sponsored young investigators as well as a safe harbor lab where scientists can conduct studies on human embryonic stem cells entirely with private funds.

The Stem Cell Program at Children's Hospital Boston

http://stemcell.childrenshospital.org/

A comprehensive, multimedia website designed to educate and inform the public about all forms of stem cells and their potential for treating human disease.

University of California San Francisco Eli and Edythe Broad Center of Regeneration Medicine and Stem Cell Research

http://stemcell.ucsf.edu/

Along with HSCI, UCSF's center was among the first in the country to focus specifically on promoting stem cell research to find new cures for disease. The site contains a newsletter as well as the latest research from the center's scientists.

For More Information on How Stem Cells Are Influencing Research on Specific Diseases

Juvenile Diabetes Research Foundation

http://www.jdrf.org

Michael J. Fox Foundation for Parkinson's Research

http://www.michaeljfox.org/index.cfm

Project A.L.S.

http://www.projectals.org/

INDEX